Crystal Reports 2008
Questions and Answers

1000 Plus Questions – Business Objects Certified Professional

By
Antonia Iroko

authorHOUSE®

AuthorHouse™ UK Ltd.
500 Avebury Boulevard
Central Milton Keynes, MK9 2BE
www.authorhouse.co.uk
Phone: 08001974150

First published by AuthorHouse   5/20/2010

ISBN: 978-1-4490-9522-2 (sc)

This book is printed on acid-free paper.

Acknowledgements

My appreciation goes to all those who supported me, both friends and family from the beginning to the end of writing this study guide and the E-books, most importantly I would like to thank God. Special thanks go to my mum for all her encouragement and support, thank you Mum and to my nieces Tobi and Tito for choosing my book cover, thank you girls! I will also like to thank all those who helped me with the previous version as this has acted as a foundation for the updated version, so a special thanks to Ade Adeoshun, Philip Gevaux, Fumilayo Dipeolu and Bijith Amat; thank you all for your time.

Credits

Author and Technical Reviewer

Antonia Iroko

Trademark Acknowledgements

Trademarked names and images materialize throughout this book; the trademark names have been used for editorial purposes only with no intention of infringing upon that trademark. This book and/or any material produced by this Author is not sponsored by, endorsed by or affiliated with Business Objects, Microsoft or Oracle. All trademarks are trademarks of their respective owners.

Limits of Liability and Disclaimer of Warranty

With the study guides and practice questions featured in this publication, we make no implications, warranties, promises or guarantees whatsoever in any manner of speaking, in whole or in part, that by joining, responding to, following, or adhering to any program or information featured anywhere in this publication or elsewhere with regards to the owner of this publication that you will be one hundred percent successful. Your success in these or any efforts depends on numerous factors. I assume no responsibility for any losses or damages resulting from your use of any link, information, or opportunity contained within this publication or within any information disclosed by the author in any form whatsoever

Contents

Introduction

Crystal Reports 2008 questions and answers study guide consists of over 1000 practice questions. This version has been updated from the BOCP for Crystal Reports – Quick reference study guide as certain features within the new version Crystal Reports 2008 have changed. I have listened to the feedback from users and arranged this book in the order of the syllabus, which is split into RDCR08201 and RDCR08301, making it easier for users to prepare for the two exams required to obtain the certification for Crystal Reports. Each chapter covers questions on the various sections of Crystal Reports utilization, functionality and development.

Crystal Reports is an advanced Business Intelligence reporting software package, which provides users with exceptional reporting functionalities; which is utilized by many companies to achieve their reporting requirements. The study guide questions will test the reader's knowledge of the functionalities within Crystal Reports and how these functions can be applied to various aspects of reporting to achieve specific goals.

Who Should Use This Book

The study guide questions are aimed at Crystal Reports 2008 Designers and Developers preparing for the certification exams (BOCP), it also acts a knowledgebase for Beginners to Advanced users. Readers are advised to use this study guide in conjunction with hands-on-practice and classroom based courses, this will give readers a greater insight into the functionalities of Crystal Reports.

Topics Covered in This Book

There are 8 chapters in this book with a total of 1007 questions. The chapters are organized as follows:

RDCR08201

Chapter 1: Create a Basic Report
Chapter 2: Customize and Format a Report
Chapter 3: Formulas

Chapter 4: Manage Reports
Chapter 5: Create an Advanced Report

RDCR08301

Chapter 6: Use Report Processing Techniques
Chapter 7: Use Subreports
Chapter 8: Create Complex Formulas and Custom Functions

Chapter 1 Create a Basic Report: In this chapter questions relating to connecting to a data source, adding tables, the design environment, creating a new report, inserting and positioning objects on a report, previewing and saving a report, applying a record selection and organizing data in a report are covered. This chapter consists of **270 questions**.

Chapter 2 Customize and Format a Report: In this chapter questions relating to formatting objects, adding graphical elements, section formatting, creating a chart and applying report templates are covered. This chapter consists of **178 questions**.

Chapter 3 Create a Formula: Questions in this chapter relate to the creation of formulas, using functions and operators which cover financial, date, maths, string, SQL expressions, control structures, variables and arrays. There are **175 questions** in this chapter.

Chapter 4 Manage Reports: This chapter covers questions on exporting a report into various formats, managing and using the workbench, managing and using the repository. There are a total of **50 questions** in this chapter.

Chapter 5 Create an Advanced Report: Questions in this chapter cover the following topics: creating a parameter, using dynamic cascading prompting, building and formatting a cross-tab report, using the running total expert, building report alerts and creating Top N reports. There are **119 questions** in this chapter.

Chapter 6 Use Report Processing: Questions in this chapter cover the following topics: multiple pass processing, using the database expert, links and join types, setting up and configuring data sources, updating data sources, processing data on the server, validating a report, distributing personalized content and using an XML transform. There are **114 questions** in this chapter.

Chapter 7 Use Subreports: This chapter covers questions on Subreports, covering linked and unlinked Subreports, On-Demand Subreports and the integration of shared variables within Subreports and linking 'un-linkable' data. There are **44 questions** in this chapter.

Chapter 8 Create Complex Formulas and Custom Functions: This chapter covers questions on creating Custom Functions; areas covered include the utilization of Custom Functions via the Extractor and Editor, using evaluation time functions, using dynamic arrays, using print state functions, using loop control structures with arrays and creating Hyperlink reports. There are **67 questions** in this chapter.

Structure of Study Guide Questions

All study guide questions are multiple-choice questions; answers are either single or multiple. Questions which require multiple answers will be indicated with *(Multiple Answers)* after the question.

Software Version

Crystal Reports 2008

Exam Required to Obtain Certification

Business Objects Certified Professional for Crystal Reports 2008 consists of two exams, **RDCR08201** and **RDCR08301**. The certification exam pass criteria is outlined in the table below, refer to the SAP Business Objects website for up to date exam requirements: http://www.vue.com/busobjects/BOCP_CR_2008_Exam_Guide.pdf

Exam	Number of Questions in Exam	Completion Time	Pass Mark
RDCR08201	62	124mins	65
RDCR08301	48	96mins	70

Exam Syllabus

Each exam certification covers the following topics. Please refer to the Business Objects website for regular exam content updates: http://www.sap.com/services/education/catalog/businessobjectstraining/index.epx

RDCR08201

Topics	Competencies Covered
Create A Basic Report	➤ Connect To A Data Sources ➤ Add Tables ➤ Describe The Design Environment ➤ Insert And Position Objects On A Report ➤ Preview And Save A Report ➤ Apply Record Selection ➤ Organize Data In A Report
Customize And Format A Report	➤ Format Objects ➤ Adding Graphical Elements ➤ Insert Fields With Pre-Built Functions ➤ Apply Section Formatting ➤ Format Data Conditionally ➤ Create A Chart ➤ Apply Record Templates
Create Formulas	➤ Create A Formula ➤ Use Functions And Operators ➤ Use Control Structures ➤ Use Variables ➤ Use Arrays
Manage Reports	➤ Export A Report ➤ Manage Reports Using The Workbench ➤ Manage Reports Using The Repository
Create An Advanced Report	➤ Creating A Parameter ➤ Use Dynamic Cascading Prompting ➤ Build And Format A Basic Cross-Tab ➤ Use The Running Total Expert ➤ Build A Report With Alerts ➤ Build A Top N Report

RDCR08301

Topics	Competencies Covered
Use Report Processing Techniques	➤ Explain The Multi-Pass Reporting Process ➤ Use The Database Expert ➤ Identify Links And Join Types ➤ Explain How To Set-Up And Configure Data Sources ➤ Set-Up And Configure Data Sources ➤ Update Reports For Database Changes ➤ Process Data On The Server ➤ Explain How To Validate Report Data ➤ Validate Report Data ➤ Distribute Personalized Content ➤ Use An Xml Transform
Use Subreports	➤ Define A Subreport ➤ Create An Unlinked Subreports ➤ Create A Linked Subreports ➤ Create An On-Demand Reports ➤ Use Shared Variables In A Subreports ➤ Use Shared Array Variables In A Subreport ➤ Link Un-Linkable Data With Subreport
Create Complex Formulas And Custom Functions	➤ Use Evaluation Time Functions ➤ Use A Dynamic Arrays ➤ Use Print State Functions ➤ Use Loop Control Structures ➤ Use Loop Control Structures With Arrays ➤ Use Custom Functions ➤ Hyperlink Reports

Booking Your Exam

Pearson VUE is the Authorized Testing Agency for booking the BOCP exams. Exams can be booked via http://www.vue.com/

RDCR08201

CHAPTER 1 - CREATE A BASIC REPORT

Creating a basic report covers questions on the creation of a datasource which is the first step in the report creation process, other areas covered include the types of data sources available to the user, adding tables to a report, understanding the design environment and report sections; inserting and positioning objects on a report, previewing and saving reports, applying record selections and organizing data in a report.

Keywords:

Data source, Administrative Tools, Control Panel, ODBC, DSN, Direct Database Drivers, Native Drivers, OLE Providers, System and User DSN, Table, Data tab, Links Tab, Available Data sources, Smart Linking, Section Name, Toolbars, Header, Footer, Details, ,Labels, Preview Panel, Convert Database Values, Grid, CASE, Page Size, Design View, Dependency checker, Unlock, Tool Tips, Server Grouping, Field Explorer, Save, Preview Icon, Record Selection, Select Expert, Formula Workshop, Group Sort Expert.

Connect to a Data Source

Q1. **Which of the following best describes the configuration of an ODBC for all users of the system?   (System Used: Windows XP)**

A. Start – Control Panel – Administrative Tools – Data sources ODBC – User DSN Tab –Add – Select database driver – enter name, description, server, set authenticity, default database, test data source connection

B. Start – Control Panel – Administrative Tools – Data sources ODBC – File DSN Tab –Add – Select database driver – enter name, description, server, set authenticity, default database, test data source connection

C. Start – Control Panel – Administrative Tools – Data sources ODBC – Report DSN Tab –Add – Select database driver – enter name, description, server, set authenticity, default database, test data source connection

D. Start – Control Panel – Administrative Tools – Data sources

ODBC - System DSN Tab –Add – Select database driver – enter name, description, server, set authenticity, default database, test data source connection

Q2.   You want to check the current version of the SQL Server ODBC driver. Which of the following apply?

A. Start – Control Panel – Administrative Tools - data sources ODBC connection - tracing tab and scroll down to the required name
B. Start – Control Panel – Administrative Tools - data sources ODBC connection - drivers tab and scroll down to the required name
C. Start – Control Panel – Administrative Tools - data sources ODBC connection - System DSN drivers tab and scroll down to the required name
D. Start – Control Panel – Administrative Tools - click the data sources ODBC connection - User DSN drivers tab and scroll down to the required name

Q3.   What are Direct Database Drivers?

A. A driver provided by Crystal
B. A driver for indirect access to Crystal
C. The driver provided by the vendor to establish Connectivity between Crystal and the vendors database
D. None of the above

Q4.   You want to connect to a Direct Database Driver, what should you do?

A. Create New Connection via the Database Expert, click more data sources and select the database required
B. Create New Connection via the view Expert, click more data sources and select the database required
C. Create New Connection via the File Expert, click more data sources and select the database required

Q5.   Which of the following information will a DSN file contain?
(Multiple Answers)

A. Driver
B. UID

C. PageTimeout
D. UserCommitSync
E. DriverId
F. SafeTransactions
G. MaxBufferSize
H. MaxScanRows

Q6. Direct Database Drivers are provided for which of the following? *(Multiple Answers)*

A. Informix
B. Lotus Domino
C. Xbase
D. Sybase, JDCB (JNDI)

Q7. Which of the following connections are available within the Database Expert? *(Multiple Answers)*

A. ODBC (RDO)
B. OLAP
C. OLE DB (ADO)
D. JDBC (JNDI)
E. ADO.NET (XML)
F. Repository
G. XML and Web Services
H. Universes
I. Salesforce.com

Q8. You receive the message 'no items found' while trying to create a Sybase connection. What does this imply?

A. You do not have access to the database
B. You need to reinstall the database
C. The appropriate database client drivers have not been installed
D. You need to reinstall Crystal

Q9. What does OLE stand for?

A. Open Link Entry
B. Object Linking and Embedding

C. OLAP Link Entry

D. Ordinary Link Enterprise

Q10. Which of the following are OLE DB providers? *(Multiple Answers)*

A. Microsoft ISAM 1.1 OLE DB Provider

B. Microsoft OLE DB Provider for DTS Packages

C. Microsoft OLE DB Provider for OLAP Services

D. Microsoft for Oracle

E. Microsoft OLE DB Provider for Sybase

F. Microsoft for SQL Server

G. Microsoft OLE DB Simple Provider

Q11. You want to create a connection to Microsoft SQL Server via an OLE DB. Which of the following apply?

H. Within the Database Expert select OLE DB, enter the details for the server, user ID, password and database and click next, property values can be edited if required and click next

I. Within the Database Expert select DB, enter the details for the server, user ID, password and database and click next, property values can be edited if required and click next

J. Within the Database Expert select the Microsoft SQL Server, enter the details for the server, user ID, password and database and click next, property values can be edited if required and click next

K. Within the Database Expert – create new connection – select OLE DB, select Microsoft OLE DB Provider for SQL Server and click next, enter the details for the server, user ID, password and database and click next, property values can be edited if required and click next

Q12. You want to create an ODBC connection, which will enable you to publish and schedule reports via Business Objects Enterprise. Which of the following should you use?

A. System DSN

B. User DSN

C. File DSN

D. Drivers

Q13. **You want to update an existing ODBC connection to reflect name changes made. What should you do?**

A. A new ODBC connection is required

B. Once an ODBC connection is created, it cannot be amended, it must be deleted

C. Within the ODBC Data Source Administrator, highlight your ODBC Connection and click the configure button, enter or select the new server name

D. None of the above

Q14. **You want to create an ODBC connection to a Microsoft Access Database. Which of the following apply?**

A. Within the ODBC Data Source Administrator, select the DSN tab - Add - select the Microsoft Access Driver from the driver list and click finish, within the OBDC Microsoft Access Setup dialog box select enter the Data source Name and description and click the select button to select the database

B. Within the ODBC Data Source Administrator, select the SND tab – Add - select the Microsoft Access Driver from the driver list and click finish, within the OBDC Microsoft Access Setup dialog box select enter the Data source Name and description and click the select button to select the database

C. Within the ODBC Data Source Administrator, select the Access tab - Add - select the Microsoft Access Driver from the driver list and click finish, within the OBDC Microsoft Access Setup dialog box select enter the Data source Name and description and click the select button to select the database

D. Within the ODBC Data Source Administrator, select the System DSN tab – Add - select the Microsoft Access Driver from the driver list and click finish, within the OBDC Microsoft Access Setup dialog box select enter the Data source Name and description and click the select button to select the database

Q15. **Which of the following defines a Silent Connection?**

A. A Silent Connection indicates all information has been provided and attempts to connect

B. Silent Connect PROMPTS for login information

C. Silent Connect attempts to connect on after information has been provided

Q16. Which if the following are connection types? *(Multiple Answers)*

A. Silent Connect
B. Simple Connect
C. Drive connect

Q17. Which of the following are the three connection types generally used?

A. ODBC/OLE DB/Direct DB drivers
B. ODBC/OLEE/SIMPLE ACCESS
C. NONE OF THE ABOVE

Q18. Which of the following are available under More Data Sources via the Database Expert? *(Multiple Answers)*

A. ACT! 3.0
B. Btrieve
C. Outlook
D. IDAPI Database DLL
E. MS IIS/Proxy Log Files
F. Field Definitions
G. Web/IIS Log Files
H. Universe

Q19. XML and Web Services are available data source within the Database Expert

A. True
B. False

Q20. You select the ADO.NET (XML), which of the following do you have to supply? *(Multiple Answers)*

A. File Path
B. Class Name
C. Class Path

D. Class Subclasss

E. Use DataSet from class

F. Use DataSet from Subclass

Q21. Which of the following are the 2 options initially available via the Database Expert? *(Multiple Answers)*

A. My Connections

B. Create New Connection

C. Connect Only

Q22. Which of the following are the 2 options available when creating a JDBC Connection? *(Multiple Answers)*

A. JDBC Connection

B. JNDI Connection

C. JJDI Connection

D. DBCJ Connection

Q23. Which of the following will appear under Direct Database Drivers? *(Multiple Answers)*

A. ACT! 30

B. Informix

C. Legacy Exchange

D. Xbase

E. Outlook

F. Sybase

Q24. Under My Connections you select properties. Which of the following will be available? (Access Database Used) *(Multiple Answers)*

A. Database DLL

B. Database Name

C. Database Type

D. Session UserID

E. System Database Path

Q25. **Which of the following are available native drivers within the Database Expert?** *(Multiple Answers)*

A. Outlook
B. xBase
C. Lotus Domino
D. Informix
E. Field Definitions

Q26. **Under which section can native drivers be located?**

A. Database Expert – Create New Connections – More Data Sources
B. Database – Create New Connections – More Data Sources
C. Database Explorer - Create New Connections – More Data Sources
D. Report Explorer - Create New Connections – More Data Sources

Add Tables

Q1: **You want to add an additional table to an existing report. Which method would you use?**

A. Choose the report expert from the database menu and select create new connection

B. Choose the file expert from the database menu and select create new connection

C. Choose the formula expert from the database menu and select create new connection

D. Choose the format expert from the database menu and select create new connection

E. Choose Database Expert from the database menu and select the database and tables from My Connections

Q2: **You add only one table to a new report you are creating. Which of the following tabs will be available within the Database Expert?**

A. Data tab Only

B. Link Tab Only

C. Data and Link Tab

Q3: **You have added tables to your report within the Database Expert. Which section will the tables appear under?**

A. Tables

B. Available Tables

C. Selected Tables

D. Tables Selected

Q4. **You have added 2 tables to your report within the Database Expert. The reports have been automatically linked; you want to switch off this feature. Which of the following apply?**

A. File – Options – Database Tab – uncheck Automatic Smart Linking under Advanced Options

B. File – Report Options – Database Tab – uncheck Automatic Smart Linking

C. Database– Options – Database Tab – uncheck Automatic Smart Linking
D. Report – Options – Database Tab – uncheck Automatic Smart Linking

Q5. **You have added several tables to your report; you are now trying to locate a particular table. How can you find the table?**

A. Right-click within the Database Expert – Links Tab and select Locate Table, highlight the table within the Locate Table dialog box
B. Right-click within the Database Expert and select Table Location, highlight the table within the Table Location dialog box
C. Right-click within the Database Expert and select Find Table Location, highlight the table within the Find Table Location dialog box
D. Right-click within the Database Expert and select Search Table Location, highlight the table within the Search Table Location dialog box

Describe the Design Environment

Q1: **Which of the following are the sections within Crystal Reports design environment?** *(Multiple Answers)*

A. Page Header
B. Page Footer
C. Detailed Footer
D. Detailed Header
E. Details
F. Report Header
G. Report Footer

Q2: **You want to display the results of the entire report in a chart. Where can you place the chart?**

A. Page Header
B. Page footer
C. Detailed Footer
D. Detailed Header
E. Details
F. Report Header
G. Report Footer

Q3: **You want your report title to appear once per report, where should you place it?**

A. Page Header
B. Page footer
C. Detailed Footer
D. Detailed Header
E. Details
F. Report Header

Q4: **You want your report title to appear once per page, where should you place it?**

A. Page Header
B. Detailed Footer
C. Detailed Header

D. Details

E. Report Header

Q5: **You want the details of all records in the database table to appear in your report. Where should you place the database fields?**

A. Page Header

B. Detailed Footer

C. Detailed Header

D. Details

E. Report Header

Q6: **You create a Grand Total Summary. Which section will it appear within the report?**

A. Page Footer

B. Details

C. Report Footer

Q7: **Which section will you place a chart or a Cross-Tab?**

A. Page Footer

B. Report Header

C. Details

Q8: **Where will you place page numbers?**

A. Page Footer

B. Report Header

C. Details

Q9: **Which order does the Crystal Report design environment appear?**

A. Report Header - Page Header - Details - Report Footer - Page Footer

B. Page Header - Report Header - Details - Report Footer - Page Footer

C. Page Header - Report Header - Details - Report Footer

D. Report Header - Details - Page Header - Report Footer - Page Footer

Q10: **What is the short section name for a Report Header?**

A. Rheader
B. R-Header
C. RH
D. R-Head

Q11: **What is the short section name for the Page Footer?**

A. PF
B. PFooter
C. P-Footer
D. Page

Q12: **What is the short section name for the Report Footer?**

A. RFooter
B. R-Footer
C. RF
D. R-Foot

Q13: **What is the short section name for the Details section?**

A. Det
B. D
C. Dets
D. DT
E. R-Det

Q14: **What is the short section name for the Page Header?**

A. PH
B. PHeader
C. P-Header
D. Page

Q15: **Which of the following are available toolbars?** *(Multiple Answers)*

A. Standard
B. Formatting

C. Insert Tools
D. Expert Tools
E. Navigation Tools
F. External Command

Q16: **Which of the following exist under the Standard toolbar?**
(Multiple Answers)

A. New Report
B. Suppress
C. open
D. Save
E. Print
F. HTML Preview

Q17: **Which of the following exist under the Formatting toolbar?**
(Multiple Answers)

A. Lock Format
B. Lock Size/Position
C. Outside Borders
D. Font color
E. Decrease Font Size
F. Increase Font Size

Q18: **Which of the following exist under the Insert Tools?**
(Multiple Answers)

A. Insert Flash Objects
B. Insert Chart
C. Insert Picture
D. Insert Cross-Tab
E. Insert File
F. Insert Report
G. Insert Text Object
H. Map

Q19: **Which of the following exist under the Expert Tools?**
(Multiple Answers)

A. Database Expert

B. Insert Chart
C. Select Expert
D. Section Expert
E. Formula Workshop
F. Template Expert
G. Highlighting
H. Group Expert
I. Group Sort Expert

Q20: **Which of the following exist under the Navigation Tools?** *(Multiple Answers)*

A. Database Expert
B. Insert Chart
C. Stop
D. Section Expert
E. Refresh
F. Template Expert
G. Show First Page
H. Show Previous Page
I. Show Next Page

Q21: **You select, File - Page Setup which of the following options will be available under the unit section?** *(Multiple Answers)*

A. Pixels
B. Inches
C. Centimeters
D. Millimetres

Q22: **You select, File - Page Setup which of the following options will be available under Orientation?** *(Multiple Answers)*

A. Portrait
B. Landscape

Q23: **You select, File - Page Setup which of the following options will be available under Margins?** *(Multiple Answers)*

A. Left

B. Right
C. Top
D. Bottom

Q24: You select, File - Page Setup which of the following options will be available under Page Options?

A. Dissociate formatting page size and printer paper size
B. Printer paper size
C. Paper size
D. Page size

Describe the Design Environment and Creating a New Report

Q1: **The Crystal Reports sample Xtreme database is designed in which format?**

A. Paradox DB
B. SQL Server
C. Oracle
D. Microsoft Access
E. Sybase

Q2: **A New Standard Report is selected which of the following will appear?**

A. Database Expert Wizard
B. File Expert Wizard
C. Report Creation Wizard
D. SQL Database Creation Wizard
E. Standard Report Creation Wizard Dialog box

Q3. **You select File New. Which of the following are the report types which can be created?** *(Multiple Answers)*

A. Standard Report
B. Blank Report
C. Cross –Tab Report
D. Mailing Label Report
E. OLAP Cube Report
F. Subreport
G. Main Report

Q4: **You select File – New and Blank Report. Which one of following dialog boxes will appear?**

A. Database Expert
B. File Expert
C. Report Creation Wizard
D. SQL Database
E. Standard Report Creation Wizard

Q5: **You select File – New – and Cross-Tab Report. Which of following dialog boxes will appear?**

A. Database Expert
B. File Expert
C. Report Creation Wizard
D. SQL Database
E. Cross-Tab Report Creation Wizard

Q6: **You select File – New and Mailing labels Report. Which of following dialog boxes will appear?**

A. Database Expert
B. File Expert
C. Report Creation Wizard
D. SQL Database
E. Mailing labels Report Creation Wizard

Q7: **Which of the following are located under the Available Data Sources?** *(Multiple Answers)*

A. My Connections
B. Universes
C. XML and Web Services
D. ODBC
E. Repository

Q8: **You have selected two tables from the Available Data Sources using the arrow (>). Which section will the tables appear under?**

A. Available Tables
B. Tables Select
C. Selected Data Tables
D. Selected Tables
E. Table Wizard

Q9: You have selected one table from the Available Data Sources and clicked next. Which of the following sections will appear within the Standard Report Creation Wizard dialog box?

A. The Links section
B. Available Fields/Fields to Display/Group By/Summarized Fields/ Filter Fields
C. Available Data Sources
D. Available Files
E. Grouping

Q10: You have selected two tables from the Available Data Sources. Which of the following sections will appear when you click next?

A. Available Fields and Fields to Display
B. Available Data Sources
C. Available Files
D. Grouping
E. Link

Q11: Which one of the following cannot be performed within the Standard Report Creation Wizard?

A. Select Data Sources
B. Select Tables
C. Select Fields
D. Summarize fields
E. Insert Subreport
F. TopN
G. Bottom
H. Charts
I. Filter Fields
J. Select Template

Q12: Within the Standard Report Creation Wizard which of the following can be performed? *(Multiple Answers)*

A. Create Links
B. Filter Fields
C. Create Groups

D. Sort fields

E. Select Report Template

Q13: **Within the Mailing labels Report Creation Wizard which of the following can be performed?** *(Multiple Answers)*

A. Linking

B. Selection of Tables

C. Selection of Fields

D. Filtering

E. Select Mailing Label Types

Q14: **Which of the following are adjustments that can be made to the labels when creating a Mailing Label Report?** *(Multiple Answers)*

A. Label Size

B. Label Structure

C. Page Margins

D. Printing Direction

E. Gaps between Labels

F. Gaps between Stamps

G. Number of Labels

Q15: **Which of the following Printing Directions are available under the printing direction section of the Mailing Labels Report Creation Wizard? ?** *(Multiple Answers)*

A. Across Then Down

B. Down Then Up

C. Up Then Down

D. Down Then Across

Q16: **Which of the following margins can be adjusted under the Page Margins section of the Mailing Labels Report Creation Wizard? ?** *(Multiple Answers)*

A. Top

B. Left

C. Right

D. Bottom

E. Forward

F. Backward

Q17: **Which of the following settings are available when setting the gaps between Labels?** *(Multiple Answers)*

A. Horizontal

B. Vertical

C. Bilateral

D. Unilateral

Q18: **Each time you refresh your report the alert dialog box appear, how can you turn off the dialog box?**

A. Select File – Report Options and uncheck Display Alerts on Refresh

B. Select File – Report Options and check Display Alerts on Refresh

C. Select Reports –Alerts and uncheck Display Alerts on Refresh

D. Select Reports –Alerts and check Display Alerts on Refresh

Q19: **The header does not appear when you drill-down within your report and go to the next page. What can you do?**

A. Select File – Format Options and check Show All Headers On Drill Down

B. Right-click the Details Section and select Section Expert and check Show All Headers On Drill Down

C. Select Report - Group Expert and check Show All Headers On Drill Down

D. Select File – Report Options and check Show All Headers On Drill Down

Q20: **How would you avoid null values from appearing on all reports created via your desktop?** *(Multiple Answers)*

A. Choose Report | Options | Reporting Tab and check Convert Database NULL Values to Default

B. Choose Format | Options | Reporting Tab and check Convert Database NULL Values to Default

C. Choose View | Options | Reporting Tab and check Convert Database NULL Values to Default

D. Choose Report | Reporting Options | Reporting Tab and check Convert Database NULL Values to Default

E. Choose File | Options | Reporting Tab and check Convert Database NULL Values to Default
F. Choose File | Options | Reporting Tab and check Convert Other NULL Values To Default

Q21: Which of the following Advanced Options can be set within the Options section of the Database Tab? *(Multiple Answers)*

A. Use indexes or Server speed
B. Perform Grouping on server
C. Database Server is Case Insensitive
D. Select Distinct Data for browsing
E. Reset the data source
F. Perform Query Asynchronously
G. Verify On First Refresh
H. Verify Stored Procedures On First Refresh
I. Verify When Database Driver Upgraded
J. Automatic Smart Linking

Q22. Which of the following will retrieve the Preview Panel within a report?

A. Edit – Group – and tick the Create Group Tree from the drop-down menu
B. File – Report Options – and check the Show Preview Panel checkbox
C. View – Group – and tick the Create Group Tree from the drop-down menu
D. Report– Group – and tick the Create Group Tree from the drop-down menu

Q23: Which of the following options can you set within the Options section of the Database Tab? *(Multiple Answers)*

A. Perform query asynchronously
B. Verify on first refresh
C. Reset the data source
D. Verify stored procedure on first refresh
E. Verify when Database driver upgraded

Q24: **You select File – Save Data with Report, however each time you open the report no data appears. What can you do to address this problem?**

A. Choose File reporting options | select the reporting tab and uncheck the box with Discard Saved Data When Loading Reports

B. Choose database | options | select the reporting tab and uncheck the box with Discard Saved Data When Loading Reports

C. Choose format | options | select the reporting tab and uncheck the box with Discard Saved Data When Loading Reports

D. Choose report | options | select the reporting tab and uncheck the box with Discard Saved Data When Loading Reports

E. Choose File | options | select the reporting tab and uncheck the box with Discard Saved Data On Open

Q25: **You want to turn the 'perform grouping on server' option ON for all reports. Which of the following applies?** *(Multiple Answers)*

A. Choose File \ Report Options \ and check the Perform Grouping On Server checkbox

B. Choose File \ Options \ Database Tab \and check the Perform Grouping On Server checkbox

C. Choose File \ Edit\ Options \ and check the Perform Grouping On Server checkbox

D. Choose File \ Database\ Options \ and check the Perform Grouping On Server checkbox

E. Choose Database \Perform Grouping On Server

Q26: **Which of the following settings apply to both Design View and Preview?** *(Multiple Answers)*

A. Rulers and Guidelines

B. Tool Tips and Section Names

C. Tool Tips and Grid

D. Short Section Names and Show Hidden Sections

E. Section Names

F. Show Hidden Sections and Page Breaks in Wide Pages

Q27. In Preview mode, you want to eliminate section descriptions, which one of the following will remove the name abbreviation?

A. File – Options –Layout Tab –Preview Section and Short Section Names
B. File – Options –Layout Tab –Preview Section and check Section Names
C. File – Report Options –Layout Tab –Preview Section and Short Section Names
D. File – Report Options –Layout Tab –Preview Section and check Section Names

Q28: You want to remove the Grid behind your text in Preview mode, which of the following apply? *(Multiple Answers)*

A. Select View - Grids and uncheck the Preview section
B. Select Report and uncheck Grids
C. Select File and uncheck Grids
D. Select File, Options, Layout Tab, under Preview uncheck Grid
E. Select File, Report Options, Layout Tab, under Preview uncheck Grid

Q29: Which setting is applicable to all reports created via the desktop?

A. Report Options
B. Options

Q30: Which of the following will retrieve a descriptive name for report sections in design mode?

A. File – Report Options –Layout Tab - Design View section and check Short Section Names
B. File – Report Options – Layout Tab - Design View section and check Section Names
C. File – Options – Layout Tab - Design View section and un-check Short Section Names
D. File – Options – Layout Tab - Design View section and check Section Names

Q31: **Which of the following settings is unique to the Design View only?**

A. Rulers and Guidelines
B. Tool Tips and Section Names
C. Tool Tips and Grid
D. Short Section Names and Show Hidden Sections
E. Section Names
F. Show Hidden Sections and Page Breaks in Wide Pages

Q32: **Which of the following settings is unique to the Preview only?** *(Multiple Answers)*

A. Rulers and Guidelines
B. Tool Tips and Section Names
C. Tool Tips and Grid
D. Short Section Names and Show Hidden Sections
E. Section Names
F. Show Hidden Sections
G. Page Breaks in Wide Pages

Q33: **Which of the following settings allows the user to set default settings for the current report only?**

A. Report Options
B. Options

Q34: **You notice fields placed in the Details sections in design mode appear as XXXXX when previewed and not as logical field names. Which of the following must be implemented to rectify this?**

A. Choose File | Report Options | under Field Options turn on the show field names check box in the field options section of the Layout tab
B. Choose File | Options | Layout tab |under Field Options section check the Show Field Names checkbox
C. Choose File | Report Options | Layout Tab| under Field Options turn on the show field names check box in the field options section of the Database tab
D. Choose File |Report Options | under Field Options turn off the

show field names check box in the field options section of the Layout tab

Q35: **How can you prevent field headings from appearing when a field is placed in the details section?**
  A. Choose Format | Options | Layout Tab | uncheck Insert Detail Field headings
  B. Choose File | Options | Layout Tab | under Field Options uncheck Insert Detail Field headings
  C. Choose Report | Options | Layout Tab | uncheck Insert Detail Field headings
  D. Choose View | Options | Layout Tab | uncheck Insert Detail Field headings

Q36. **The group heading does not appear when a group is inserted within your report, how can you change this?**

  A. Select File – Report Options – Layout Tab – and check the Insert Group Name with Group checkbox
  B. Select File – Options – Layout Tab –under Field Option and check the Insert Group Name with Group checkbox
  C. Select File – Report Options – Reporting Tab – and check the Insert Group Name with Group checkbox
  D. Select File – Options – Reporting Tab – and check the Insert Group Name with Group checkbox

Q37. **You want your report to be opened with a page size of 100%, which of the following apply?**

  A. File – Options – Change Preview Page Starts with Fit Page to Full Size
  B. File – Options – Layout Tab and check Pages Start with Full Size (100%)
  C. File – Report Options – Change Preview Page Starts with Fit Page to Full Width
  D. File – Options – Change Preview Page Starts with Fit Page to Full Width

Q38. **You want to change the database driver from being CASE sensitive. How will you implement this?** *(Multiple Answers)*

A. From the Menu Bar, select File | options | Database Tab | under the advanced options check the database server is case-insensitive
B. From the Menu Bar, select File | report options and check the database server is case-insensitive
C. Database Expert - Right-click Database Connection – Select Options and check database server is case-insensitive
D. Database Expert - Right-click Database Connection – Database Link Options and uncheck database server is case-insensitive

Q39. **You want to view only tables owned by Antonia when creating any reports.   Which of the following apply?** *(Multiple Answers)*

A. Select File, Report Options, Database Tab, under the Data Explorer section enter Anto% under Owner LIKE: (%,_)
B. Select File, Options, Database Tab, under the Data Explorer section enter Anto% under Owner LIKE: (%,_)
C. Select Database, Database Expert from the Menu Bar within your connection, right-click tables, select options from the drop-down menu, under the Data Explorer section enter Anto% under Owner LIKE: (%,_)
D. Select Report, Database Expert from the Menu Bar within your connection, right-click tables, select options from the drop-down menu, under the Data Explorer section enter Anto under Owner LIKE: (%,_)

Q40. **You have applied a Tool Tip to your report, however each time you place the cursor over the text, in preview mode, the tool-tip does not appear. Which of the following will resolve the issue?**

A. File –Report Options –Layout Tab – and check the Tool Tips check box under preview
B. View –Layout Tab – and check the Tool Tips check box under preview
C. Format –Options –Layout Tab – and check the Tool Tips check box under preview

29

D. File –Options –Layout Tab – and check the Tool Tips check box under Preview

Q41. **You are currently working with tables from the Finance department. Which start with Fin_. You want to see only these tables when creating any report.** *(Multiple Answers)*

A. Select File, Report Options, Database Tab, under the Data Explorer section enter Fin% under Table name LIKE: (%,_)
B. Select File, Options, Database Tab, under the Data Explorer section enter Fin% under Table name LIKE: (%,_)
C. Select Database, Database Expert from the Menu Bar within your connection, right-click tables, select options from the drop-down menu, under the Data Explorer section enter Fin% under Table name LIKE: (%,_)
D. Select Report, Database Expert from the Menu Bar within your connection, right-click tables, select options from the drop-down menu, under the Data Explorer section enter Fin% under Table name LIKE: (%,_)

Q42. **How can you ensure all stored procedures used for report creation are being verified?**

A. Choose File \ Options within the Advanced Options check the verify stored procedures on 1st refresh
B. Choose File \ Report Expert within the Advanced Options check the verify stored procedures on 1st refresh
C. Choose View \ Report Options within the Advanced Options check the verify stored procedures on 1st refresh
D. Choose File \ Options\Database Tab\ Advanced Options\ check the verify stored procedures on First refresh

Q43. **You want to view only the System Tables within the Database Expert, What should you do?**

A. Right click the database from the Database Expert select options from the drop-down menu, within the Data Explorer section, uncheck table, Stored Procedures, Synonyms and views and check the System Tables' checkbox.
B. Right click the database from the SQL Expert Data Explorer

select options, from the drop-down menu, uncheck table, Stored Procedures, Synonyms and views and check the systems tables checkbox

C. Right click the database from the Database Expert Data Explorer select report options, from the drop-down menu, uncheck table, Stored Procedures, Synonyms and views and check the System Tables' checkbox.

D. Right click the database from the SQL Expert Data Explorer select report options, from the drop-down menu, uncheck table, Stored Procedures, Synonyms and views and check the systems tables checkbox

Q44. **You want to list tables and fields by name, as they appear in the database. Which of the following methods should you implement?**

A. Select File, Report Options, and Database Tab, then select Show Name

B. Select File, Smart Tag Options, and Database Tab, then select Show Name

C. Select File – Options – Database Tab – under the Tables and Fields section, select the Show Name Button

D. Select File, Report Expert Options, and Database Tab, then select Show Name

Q45. **Which of the following field data types can be set within the Options Fields Tab?** *(Multiple Answers)*

A. String
B. Number
C. Text
D. Field
E. Currency
F. Date
G. Time
H. Date and Time
I. Boolean

Q46. **Which of the following Report-Checking Options are available via the Dependency Checker Options Tab?** *(Multiple Answers)*

A. Compile statistics
B. Compile Formulas
C. Compile Custom Functions
D. Validate Files
E. Validate Report
F. Validate Hyperlinks URLs
G. Verify Database
H. Compile SQL Expressions

Q47. **Which of the following Multi-Report Checking Options are available via the Dependency Checker Options Tab?** *(Multiple Answers)*

A. Check Files
B. Check Repository Custom Functions
C. Check Repository objects (text and bitmap)
D. Check reports part Hyperlinks
E. Check sub files which are imported
F. Check Subreports which are re-imported on open
G. Check Subreports which are re-imported on close

Q48. **You have hidden the details section in your report. However, in the design view, this section still shows, with a grey section behind it. How do you completely hide this section in the design view?**

A. From the Section Expert, select detail section and tick hide drill-down OK
B. From the Select Expert, select detail section and tick hide drill-down OK
C. Select File | Options | select the Layout Tab and uncheck Show Hidden Sections
D. From the report expert, select detail section and tick hide drill-down OK
E. Select File | Options | select the Layout Tab and check Show Hidden Sections

Q49. **You want to prevent the Guidelines, Rulers, ToolTips, Section Names and Page Breaks from appearing in Preview mode, what action is required?**

A. Select View and uncheck guidelines, Rulers, ToolTips, section names and page breaks appear

B. Select Report and uncheck guidelines, Rulers, ToolTips, section names and page breaks appear

C. Select File and uncheck guidelines, Rulers, ToolTips, section names and page breaks appear

D. Select File, Options, Layout Tab, under Preview uncheck guidelines, Rulers, ToolTips, section names and page breaks appear

E. Select File, Report Options, Layout Tab, under Preview uncheck guidelines, Rulers, ToolTips, section names and page breaks appear

Q50. **Several users have called to say they cannot see the Stored Procedures in the database when trying to create reports via the ODBC data source connection. What advise will you give the users?** *(Multiple Answers)*

A. Select File – Options – Database Tab – from the Data Explorer section – check the Stored Procedure checkbox

B. Right-click the field within the Database Expert and select options from the drop-down list and check the show Stored Procedures checkbox

C. Right-click View Expert within the Database Expert and select options from the drop-down list and check the show Stored Procedures checkbox

D. Right-click the database within the Database Expert, select options from the drop-down list and check the show Stored Procedures checkbox under Data Explorer

Q51. **You notice the tool tip appears each time you place the cursor over a field (In Preview Mode) in your report; you want to turn off the Tool Tip. Which of the following apply?**

A. File – Options – Layout Tab – uncheck Tool Tips checkbox in the preview section

B. File – Options – Reporting Tab – check Tool Tips checkbox in the preview section

C. File – Report Options – Reporting Tab – check Tool Tips checkbox in the preview section
D. File – Options – Formula Editor Tab – check Tool Tips checkbox in the preview section

Q52. How can you customize the color of the comments when creating a formula?

A. File, Report Options, Formula Editor tab, highlight comment under color element section and change foreground to required color
B. File, Report Options, Formula Editor tab, highlight comment under the Default Formula Languages section and change foreground to required color
C. File, Options, Formula Editor tab, highlight comment under the color element section and change foreground to required color
D. File, Options, Formula Editor tab, highlight comment under the Default Formula Languages section and change foreground to required color

Q53. Which option setting is universal for the desktop it is set on?

A. Report Options
B. Options
C. File Options
D. Report Settings

Q54. Which of the following formula elements can you change?
(Multiple Answers)

A. Comment
B. Text Selection
C. Formula
D. Custom Function
E. Report Data Type
F. Keyword
G. Text

Q55. Where can you set the Grid size for your reports?

A. File – Options – Fields Tab – under the Grid options section, set Grid Size

B. File – Options – Layout Tab – under the Grid options section, Grid Size

C. File – Options – Font Tab – under the Grid options section, Grid Size

D. File – Options – Reporting Tab – under the Grid options section, Grid Size

Q56. **You want to set all fields within your report to the same font size. Which of the following applies?**

A. File – Report Options – Fields Tab, select field type tab within the field format section and format field types as required

B. File – Options – Font Tab and format field types as required

C. Report -Options – Fields Tab, select field type tab and format field type as required

D. Database – Report Options – Fields Tab, select field type tab and format field type as required

Q57. **Which platforms are available under Smart Tag and HTML Preview?** *(Multiple Answers)*

A. .Net

B. Java

C. Windows XP

D. Unix

Q58. **You want to ensure that each time the main report is opened the Subreport is re-imported. What should you do?**

A. Select |File| report options| reporting TAB| check the re-import Subreport when opening reports.

B. Select |File| Edit options | reporting TAB | check the re-import Subreport when opening reports.

C. Select |Database options | reporting TAB | check the re-import Subreport when opening reports.

D. File | Options | Reporting TAB and check the re-import Subreport on Open.

Q59. You want to add the following information to your report: Author, Title, Keywords, Title, Subject, and Comments. Which of the following apply?

A. File | Report Options | as Enter details required
B. File | Options |as Enter details required
C. Report | Report Options |as Enter details required
D. File | Summary Info | as Enter details required

Q60. Which of the following will activate automatic linking within the Database Expert Links Tab?

A. File - Report Options – check the Automatic Link checkbox
B. File – Options – Check the Automatic Link checkbox
C. File –Report Options – Database Tab – Advanced Options and check the Automatic Smart Linking checkbox
D. File –Options – Database Tab – Advanced Options sections and check the Automatic Smart Linking checkbox

Q61. You want to check the report attributes: the total time taken to edit a report; the last person to save the report; the revision number; last printed; created; and saved. Where can this information be found?

A. File |Summary Info | Summary Tab
B. File | Summary Info | Info Tab
C. File | Summary Info | Statistics Tab
D. File |Summary Info | Edit Tab

Q62. You want to make your report available via the template directory. Which of the following apply?

A. Choose Report | Summary Info | Check the Save Preview Picture checkbox
B. Choose File | Summary Info | Check the Save Preview Picture checkbox
C. Choose File | Report Options | Summary Info | Check the Save Preview Picture checkbox
D. Choose File | Options | Check the Save Preview Picture checkbox

Q63. **You do not want the Preview Panel to appear in any report created via your desktop. Which of the following apply?**

A. File – Options – Fields Tab – uncheck Display Group Tree
B. File – Report Options – Fields Tab – uncheck Display Preview Panel
C. File – Options – Layout Tab – uncheck Display Preview Panel
D. File – Report Options – Layout Tab – uncheck Display Group Tree
E. File – Options – Layout Tab – check Display Group Tree

Q64. **You want to arrange tables in the Field Explorer in an alphabetical order. Which of the following apply?** *(Multiple Answers)*

A. File - Report Options - and check the Sort Tables Alphabetically checkbox
B. File – Options – Database Tab- and under Tables and Fields check the Sort Tables Alphabetically checkbox
C. Database – Database Expert - Right-click the database and select options from the drop-down list and check the Sort Tables Alphabetically checkbox under Tables and Fields
D. File - Options - and check the Sort Tables Alphabetically checkbox

Q65. **You want to prevent users from modifying the sales report you have created. Which of the following apply?**

A. Select File – Options and check the Save Lock Report Design checkbox, enter new password and confirm password and click OK – OK
B. Select File – Report Options and check the Save Lock Report Design checkbox, enter new password and confirm password and click OK – OK
C. Select Report – Report Options and check the Save Lock Report Design checkbox, enter new password and confirm password and click OK – OK
D. Select Database – Report Options and check the Save Lock Report Design checkbox, enter new password and confirm password and click OK – OK

Q66.   Using Save As can modify a locked report?

A. True
B. False

Q67.   How would you unlock your report?

A. Select File – Options –and uncheck the Save Lock Report Design, you will be prompted for a New Password, enter your password, Confirm Password and click ok
B. Select Report – Report Options –and tick the Save Lock Report Design, you will be prompted for a New Password, enter your password, Confirm Password and click ok
C. Select File – Report Options –and uncheck the Save Lock Report Design, you will be prompted your Password, enter your password and click ok and save
D. Select Report - Options –and tick the Save Lock Report Design, you will be prompted for a New Password, enter your password, Confirm Password and click ok

Q68.   You have checked the following checkboxes 'Perform Grouping on Server' and 'Use Indexes or Server for Speed' within the Report Options section. Which of the following will be automatically greyed out?

A. Database Server is Case-Insensitive
B. Perform Query Asynchronously
C. Convert Other NULL Values to Default
D. Convert Database NULL Values to Default

Q69.   You try to format fields within your report, all formatting options are grayed out and you cannot move any fields within your report. Which of the following Report Options has been set?

A. Save Lock Report Design
B. Read-only

Q70. **You want your report to be saved after every two minutes. Which of the following settings apply?**

A. File – Report Options - check the Autosave Reports After checkbox and enter 2 minutes
B. File – Options - check the Autosave Reports After checkbox and enter 2 minutes
C. File – Options - Reporting Tab - and check the Autosave Reports After checkbox and enter 2 minutes
D. File – Report Options - Reporting Tab and check the Autosave Reports After checkbox and enter 2 minutes

Q71. **You want to ensure all Repository based objects within your reports are updated when the report is opened. Which of the following settings apply?**

A. File – Report Options - under Enterprise Settings check the Update Enterprise Report Properties on Save
B. File - Options - under Enterprise Settings check the Update Enterprise Report Properties on Save
C. File – Options - Database Tab - under Enterprise Settings check the Update Enterprise Report Properties on Save
D. File – Options - Reporting Tab - under Enterprise Settings check the Update Connected Repository Objects on Open checkbox

Q72. **You select Start – All programs – Crystal Reports 2008. Which of the following will appear?**

A. Crystal Reports Start Page
B. A Blank Crystal Report
C. The Database Expert
D. The Report Wizard

Q73. **Which of the following is available under the Start A New Report section?** *(Multiple Answers)*

A. Blank report wizard
B. Cross-Tab report wizard
C. Mailing label report wizard
D. OLAP Cube Report Wizard
E. Standard Report

Q74. Which of the following Tabs are available on the start page? *(Multiple Answers)*

A. Highlights
B. Downloads
C. Developer
D. IT Professional
E. Report Designer

Q75. You want to suppress all blank records? Which of the following apply?

A. File – Report Options – Suppress Printing if no records
B. File – Options – Suppress Printing if no records
C. File – Suppress Printing if no records
D. Report – Report Options – Suppress Printing if no records

Q76. What does the Status Bar display? *(Multiple Answers)*

A. Date and Time
B. Records
C. Zoom to whole page
D. Zoom to page width

Q77. You select Zoom level within the status bar. Which of the following can be applied? *(Multiple Answers)*

A. Magnification Factor
B. Fit One Dimension
C. Fit Whole Page
D. Fit Part Page

Q78. Which of the following are free product add-ons? *(Multiple Answers)*

A. Sample Reports + Database
B. Runtime Packages
C. Crystal Reports Eclipse
D. .NET SDK documentation
E. JAVA SDK documentation
F. Crystal Reports Viewer 2008
G. Data Direct ODBC Drivers

H. ESRI GIS Map Viewer

I. Crystalreports.com report Sharing service

Q79. **Which of the following are the tabs available on the Start Page?** *(Multiple Answers)*

A. Highlights
B. Download
C. Developer
D. IT Professional
E. Report Designer

Insert and Position Objects on a Report

Q1. **You can insert database fields from which of the following?**

    A. Field Explorer
    B. Report Explorer
    C. Workbench

Q2. **Which of the following objects can be inserted onto a report?** *(Multiple Answers)*

    A. Database Fields
    B. Lines
    C. Boxes
    D. Formula Fields
    E. Loops

Q3. **Which of the following can be inserted on to the report from the Field Explorer?** *(Multiple Answers)*

    A. Database Fields
    B. Special Fields
    C. Formula Fields
    D. SQL Expression Fields
    E. Parameter Fields
    F. Group Name Fields
    G. Running Total Fields

Q4. **Which method can be used to insert a field onto a report from the Field Explorer?** *(Multiple Answers)*

    A. Right-click the field and select Insert to Report
    B. Click the field and drag it on to the report
    C. Double-click the field and Insert on to Report

Q5. **The Report Explorer displays an outline of the object on the report.**

    A. True
    B. False

Q6. Objects within a report can be modified within the Report Explorer?

A. True
B. False

Q7. Which of the following options will be available when a field is clicked within the Report Explorer? *(Multiple Answers)*

A. Format Field
B. Format Painter
C. Select Expert Group
D. Select Expert Record
E. Select Expert Saved Data

Q8. Which of the following objects are available within the Repository Explorer? *(Multiple Answers)*

A. Commands
B. Custom functions
C. Text objects
D. Bitmaps
E. Business Views
F. List of Values

Q9. Enterprise Connectivity is required to view objects within the Repository Explorer.

A. True
B. False

Q10. Which of the following types of OLE objects can be inserted into a report? *(Multiple Answers)*

A. Paintbrush Picture
B. Wave Sound
C. Video Clip
D. Static Report

Q11. You want to insert a field from the Field Explorer onto your report. Which of the following methods apply? *(Multiple Answers)*

A. Double-click the report from within the Field Explorer and drag to the desired position on the report
B. Click the field from within the Field Explorer and holding down the mouse drag to the desired position on the report and release the mouse.
C. Right-click the report from within the Field Explorer and select insert onto report from the drop down list
D. Right-click the field from within the Field Explorer and select Insert to report, drag field to the desired position and release mouse

Q12. You want to move the Customer's name, number and order date to the group footer, you select all the fields at once and right-click the customer's name field, which of the following options for moving these fields will appear in the drop down list? *(Multiple Answers)*

A. Paste
B. Cut
C. Delete
D. Edit
E. Copy

Preview and Save a Report

Q1. **Preview can be activated using which of the following?**
(Multiple Answers)

A. Report – Refresh Report Data
B. View - HTML Preview (Only works with Enterprise Connectivity)
C. View - Print Preview
D. F5
E. View - Preview Sample

Q2. **Which of the following Icons can be used to refresh a report?**

A.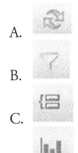

B.

C.

D.

Q3. **To access the preview options select which of the following?**

A. From the Menu bar select File - Preview
B. From the Menu bar select Report - Preview
C. From the Menu bar select View - Preview
D. From the Menu bar select Database - Preview

Q4. **You want to preview the first 100 records of your report. Which of the following methods apply?**

A. Select file, Preview, from the Preview Sample dialog box select the First button and enter 100.
B. Select View, Preview Sample, from the Preview Sample dialog box select the First button and enter 100.

C. Select Report, Preview, from the Preview Sample dialog box select the First button and enter 100.

D. Select refresh and Preview, from the Preview Sample dialog box select the First button and enter 100.

Q5.   Previewing a web view of your report involves which of the following methods?

A. From the Menu Bar, select View, XML Preview
B. From the Menu Bar, select View, SXML Preview
C. From the Menu Bar, select View, DHTML Preview
D. From the Menu Bar, select View, HTML Preview

Q6.   The Preview Icon is greyed out. What does this indicate?

A. The report requires refreshing
B. The Report has failed to preview
C. The Report is not connected to the database
D. The Report is already in preview mode

Q7.   In preview mode you want to magnify the view. Which of the following apply? *(Multiple Answers)*

A. Select windows from the Menu Bar and select zoom, select the magnification factor and enter the required percentage
B. Select report from the Menu Bar and select zoom, select the magnification factor and enter the required percentage
C. From the status bar slide the zoom control to the appropriate zoom percentage
D. Select view from the Menu Bar and select zoom, select the magnification factor and enter the required percentage

Q8.   Which of the following icons represent the Print Preview?

A.

B.

C.

D.

Q9. Which of the following icons represent the HTML Preview?

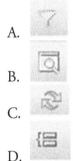

A.

B.

C.

D.

Q10. The preview icon is greyed out. You want to reactivate the preview button. Which one of the following applies?

A. From the Menu Bar, select View, Delete Current View
B. From the Menu Bar, select View, Current View Shut
C. From the Menu Bar, select View, Close Current View
D. From the Menu Bar, select View, Close View
E. Click the cross beside the Design Icon

Q11. You have clicked the Print Preview button. Which of the following sections will be visible? *(Multiple Answers)*

A. The Preview Panel
B. The Report in preview mode
C. The Field Explorer
D. The Repository Explorer

Q12. Where can you view recent reports?

A. Front Page
B. Last Page
C. View Page
D. Start Page

Q13. **Which of the following are available from the Start Page?**
(Multiple Answers)

A. Sample Reports and Databases
B. Data Direct ODBC Drivers
C. .NET SDK documentation
D. JAVA SDK documentation

Q14. **You want to check for current updates to Crystal Reports. Which of the following apply?**

A. Front Page\Check for updates
B. Last Page\Check for updates
C. View Page\Check for updates
D. Start Page\Download\Critical updates

Q15. **You want to save a new report. Which of the following apply?** *(Multiple Answers)*

A. Select Report and Save, the Save As dialog box will appear, select the folder you want to save the report into and give the report a name and click the save button
B. Select View and Save, the Save As dialog box will appear, select the folder you want to save the report into and give the report a name and click the save button
C. Select File and Save, the Save As dialog box will appear, select the folder you want to save the report into (Save in) and give the report a name (File Name) and click the save button

D. Select the Save icon , the Save As dialog box will appear, select the folder you want to save the report into and give the report a name and click the save button

Q16. **You want other users to see the last data you saved with your report. Which of the following apply?**

A. Select Report and Save Data with Report when saving the report
B. Select Database and Save Data with Report when saving the report
C. Select File and Save Data with Report when saving the report
D. Select View and Save Data with Report when saving the report

Q17. You want to save an existing report into a different folder. Which of the following apply?

A. Select File and Save As, the Save As dialog box will appear, select the folder you want to save the report into and give the report a name and click the save button

B. Select the Save icon, the Save As dialog box will appear, select the folder you want to save the report into and give the report a name and click the save button

C. Select Report and Save As, the Save As dialog box will appear, select the folder you want to save the report into and give the report a name and click the save button

D. Select View and Save As, the Save As dialog box will appear, select the folder you want to save the report into and give the report a name and click the save button

Q18. You have opened several reports and all reports appear with their own tabs. You want to close the Cross-Tab report without closing all other opened reports. How would you do this? *(Multiple Answers)*

A. Select the Cross-Tab report tab and close

B. Select the Cross-Tab report tab and click the cross, a dialog box will appear prompting you to save, (Yes, No or Cancel)

C. From the Menu Bar, select View, Report and close Report

D. Select the Cross-Tab report tab , from the Menu Bar, select File and close, a dialog box will appear prompting you to save, (Yes, No or Cancel)

Q19. You want to save a Subreport. Which of the following apply?

A. Highlight the Subreport and select File - Save Subreport As, the Save As dialog box will appear, select the folder you want to save the report into and give the report a name and click the save button

B. Highlight the Subreport and select Report, Save Subreport As, the Save As dialog box will appear, select the folder you want to save the report into and give the report a name and click the save button

C. Highlight the Subreport and select View, Save Subreport As, the Save As dialog box will appear, select the folder you want to save the report into and give the report a name and click the save button

D. Select File save as the Save As dialog box will appear, select the folder you want to save the report into and give the report a name and click the save button

Q20. You have moved a field to a different section within your report and you click print preview. Which of the following will take place?

A. Crystal Reports will refresh the report from the Database
B. Crystal Reports will use the data saved within the report when you preview the report
C. Crystal Reports will prompt you with a refresh option
D. Crystal Reports will inform you of the recent change

Q21. You have changed the Record Selection of your report. What will happen?

A. Crystal Reports will use the data saved with report when you preview the report
B. Crystal Reports will prompt you with a refresh option
C. Crystal Reports will inform you of the recent change
D. Crystal Reports will re-query the database and it will do so without prompting you
E. Crystal Reports will give you an option to refresh the report by using saved data or refresh the report by re-querying the database

Q22. You have created a report and saved the data with the report. Which of the following apply?

A. Saved Data is the current data in the database based on the time the report has been opened
B. Saved Data is the Crystal Administrators last data run
C. Saved Data is data based on the last time the report was run
D. None of the above

Q23. When a report is refreshed which of the following occurs?

A. Saved data with the report is showed
B. The report will fail if saved data is not ticked
C. The report queries the database
D. None of the above

Q24. You open an existing report and it opens in preview mode. Which of the following apply?

A. The report is invalid
B. Report was saved with data
C. The database no longer exist
D. None of the above

Q35. You open an existing report saved with data and apply a filter. Which of the following will occur?

A. The report will filter down the existing records within the report and return only data applicable to the current record selection
B. The report query the database and return only data applicable to the current record selection
C. The report will return an error message reading' use Refresh data only'
D. None of the above

Q36. Which of the following will appear in the Preview Panel?
(Multiple Answers)

A. Groups
B. Parameters
C. Reports
D. Find

Q37. Within the Preview Panel the parameter selection can be changed?

A. True
B. False

Q38. Within the Preview Panel the parameter selection can be changed and the changes need to be applied by clicking the Apply changes Icon?

A. True
B. False

Q39. **Which of the following options exists under the Preview Panel - Parameter section?** *(Multiple Answers)*

A. Prompt for Parameters
B. Remove Value
C. Revert All Changes
D. Apply Changes
E. None of the above

Q40. **Which of the following options exists under the Preview Panel - Find section?** *(Multiple Answers)*

A. Match Case
B. Match Number
C. Match Format
D. Match whole word only

Q41. **You notice the Preview Panel – Parameter section does not contain the parameters which exist within your report, which of the following explains this?**

A. The Preview Panel need to be refreshed
B. The Preview Panel should be deleted and redesigned
C. The Show on (Viewer) Panel was set to 'Do not Show' when creating the parameter
D. None of the above

Q42. **Which of the following Preview Panel options exist within the Create New Parameter dialog box?** *(Multiple Answers)*

A. Show
B. Do not Show
C. Editable
D. Read Only

Q43. **The Preview Panel consists of which of the following?** *(Multiple Answers)*

A. Groups
B. Parameter
C. Find
D. Chart Panel

Q44. **You want to save your existing report to Business Objects Enterprise, which methods apply?** *(Multiple Answers)*

A. File –Save As – Enterprise – Log in and save to required folder
B. Within the Repository Explorer save the report
C. View Report – Save As – Enterprise – Log in and save to required folder
D. Format – Report – Save As – Enterprise – Log in and save to required folder

Apply Record Selection

Q1. **You want to create a report for the top 100 sales return per city. Which of the following must you implement?**

A. Insert a group based on the city field and a summary sum of sales based on the group created; from the Menu Bar select report and group expert and select top 100 based on the sum created.

B. Insert a group based on the city field and a summary sum of sales based on the group created; from the Menu Bar select report and record sort expert and select top 100 based on the sum created.

C. Insert a group based on the city field and a summary sum of sales based on the group created, from the Menu Bar select report and Group Sort Expert and select Top N 100 based on the sum created.

D. Insert a group based on the city field and a summary sum of sales based on the group created, from the Menu Bar select report and Section Expert and select Top N where N is 100 based on the sum of sales created.

Q2. **You want to apply a date range filter to your report. Which of the following apply?** *(Multiple Answers)*

A. Is Between
B. Is in Range
C. Is greater than or equal
D. Is < =
E. Is Not Between

Q3. **You have received an urgent request to filter the sales report to show sales records in the months of July, August and September of the current year. Which of the following methods will enable you to achieve this without creating a complex formula?**

A. From the Menu select Report| Select Expert| and select New| select the Date field from the Choose Field Dialog box and click ok| select 'is in the period' | select Calendar1stQuarter

B. From the Menu select Report| Select Expert| and select New| select the Date field from the Choose Field Dialog box and click ok| select 'is in the period' | select Calendar1stQtr

C. From the Menu select Report| Select Expert| Record |and select New| select the Date field from the Choose Field Dialog box and click ok| select 'is in the period' | select Calendar3rdQtr

D. From the Menu select Report| Select Expert| and select New| select the Date field from the Choose Field Dialog box and click ok| select 'is in the period' | select Calendar4thQuarter

Q4. You select a string field within the Select Expert Record and leave the 'Is Equal' section Blank. Which of the following will occur?

A. {TableName.FieldName} = " "
B. An error message will be produced
C. {TableName.FieldName} <> " "
D. {TableName.FieldName}

Q5. You have used 'Is One Of ' in the Select Expert this will allow one of the following

A. Only two items to be specified
B. Numeric values only
C. Discrete values only
D. Multiple value selection

Q6. You want to extract data for the current year only. Which of the following Select Expert options apply?

A. Is One Of
B. Calendar4thQuarter
C. YearToDate
D. Is greater than

Q7. You have created a sales report and would like to extract data for the First Quarter of the Calendar Year. What is the best method to follow?

A. Right-click the date field, | from the drop down menu select format field and apply your formula for the CalendarFirstQtr.
B. Choose Report | Section Expert | apply Calendar1stQtr
C. Highlight the date field, choose report | Select Expert | Record| from the drop down menu select is in period | Calendar1stQtr.
D. Choose File | options and apply the CalendarFirstQtr formula

Q8. **Which two are not valid date functions within the Select Expert?** *(Multiple Answers)*

A. Last9Days
B. Last7Days
C. Last4WeeksToFri
D. Last4WeeksToSun
E. LastFullMonth
F. LastFullWeek
G. All DatesToToday

Q9. **The 'Is like' operator will only work with a string**

A. True
B. False

Q10. **You have created a report and would like to extract records that contain the partial letters"RDCR08" from the EXAM_ID field. Which of the following apply?**

A. {Exams.Exam_ID} Startswith 'RDCR'
B. {Exams.Exam_ID} Is between
C. {Exams.Exam_ID} like "*RDCR08*"
D. {Exams.Exam_ID} Is equal to 'RDCR08'

Q11. **You select a string and do not specify a value in the 'Is equal' section of the Select Expert; this will produce an error message**

A. True
B. False

Q12. **You have created a sales report and would like to extract data for the current month. Select the two options which apply?** *(Multiple Answers)*

A. Right-click the date field, | from the drop down menu select format field and apply your formula for the YearToDate.
B. Choose Report | Section Expert | apply YearToDate
C. Highlight the date field, choose Report | Select Expert |Record| from the drop down menu select 'is in the period' | MonthToDate.

D. ({TableName.SalesDate}) in monthtodate

E. month({TableName.SalesDate}) in monthtodate

Q13. **You have created a sales report and would like to extract data based on the Sales Date for the previous complete week. What is the best method to follow?**

A. Week({Sales.Sales Date}) in Week – 1

B. {Sales.Sales Date} in LastFullWeek

C. {Sales.Sales Date} in LastFullWeek – 7

D. {Sales.Sales Date} in LastFullWeek + 7

Q14. **Which of the following are not valid date options when using the Select Expert Tool 'Is In The Period'?** *(Multiple Answers)*

A. WeekToDateFromMonday

B. WeekToDateFromToday

C. WeekToDateFromSun

D. WeekToDateFromSat

E. MonthToDate

F. YearToDate

Q15. **Which one of the following is not available for a Date Format in the Select Expert when using 'Is In The Period'?**

A. DatesToYesterday

B. AllDatesFromToday

C. AllDatesFromTomorrow

D. Aged0To30Days

E. Aged31To60Days

F. Aged61To90Days

G. Aged0To31Days

Q16. **You have received an urgent request to filter the sales report to show sales records in the last three months of the current year. Which of the following methods will enable you to achieve this without creating a complex formula?**

A. From the Menu select Report| Select Expert| Record | and select New| select the Date field from the Choose Field Dialog box and click ok| select 'is in the period' | select Calendar1stQuarter

B. From the Menu select Report| Select Expert | Record | and select New| select the Date field from the Choose Field Dialog box and click ok| select 'is in the period' | select Calendar4thQtr

C. From the Menu select Report| Select Expert | Record | and select New| select the Date field from the Choose Field Dialog box and click ok| select 'is in the period' | select Month4thQuarter

D. From the Menu select Report| Select Expert | Record | and select New| select the Date field from the Choose Field Dialog box and click ok| select 'is in the period' | select Calendar2stQuarter

Q17. You want to extract records where the candidate's names begin with 'Anto' or 'Iro'. Which of the following must you use? (Multiple Answers)

A. Is equal to
B. Is One Of
C. Is Between
D. Starts with
E. Like '**'
F. Left({Client.ClientName},2) in ["Anto","Iro"]

Q18. Is Any Value has been selected, this will

G. Eliminate any other selection criteria you have applied to the report
H. Produce an error message
I. Only work if you leave the section blank
J. Include all records

Q19. What is the Default option within the Select Expert?

A. Is equal to
B. Is Any Value
C. Is in
D. Is Like

Q20. You want to type in a formula directly into Records Selection Formula Editor. Which of the following apply?

A. View – Select Expert – Formula – Show Formula – Formula Editor

B. Report – Select Expert – Record – Formula Editor

C. Format – Select Expert – Formula – Show Formula – Formula Editor

D. Edit – Select Expert – Formula – Show Formula – Formula Editor

Q21. **You want to extract the following data. Which of the following will apply?** *(Multiple Answers)*

Product Type Name = Server and
Product Class = 2008 and
Quantity Purchased is less than or equal to 50 and
Year Ordered is in the Current Year

A. {Product.Product Type Name} = "Server" and {Product.Product Class} = "2008" and {Orders_Info.Quantity} <= 50 and {Orders.Required Date} in YearToDate

B. {Product.Product Type Name} = "Server" and {Product.Product Class} = "2008" and {Orders_Info.Quantity} <= 50 and Year({Orders.Required Date}) = Year(CurrentDate)

C. {Product.Product Type Name} = "Server" and {Product.Product Class} = "2008" and {Orders_Info.Quantity} in 0 to 50 and Year({Orders.Required Date}) = Year(CurrentDate)

D. {Product.Product Type Name} = "Server" and {Product.Product Class} = "2008" and {Orders_Info.Quantity} < 50 and Year({Orders.Required Date}) = Year(CurrentDate)

Q22. **You want to add a Group Selection criteria to your report. Which two apply?**

A. Report – Select Expert – Group – Select Expert Group dialog box - Formula Editor and enter your criteria

B. From the Select Expert, click the Record Selection radio button, select formula editor and enter your criteria

C. From the Select Expert, click the records tab, select formula editor and enter your criteria

D. From the select Section Expert, click the Group Selection radio button, select formula editor and enter your criteria

Q23. You have created a sort order on the following fields {Exam.ExamName} and { Exam.ExamCity} and would like to reverse the order. You can reverse the order of the fields from the Record Sort Expert by selecting Report – Record Sort Expert and using the arrow to sort the fields

A. True
B. False

Q24. Which of the following are Select Expert options? *(Multiple Answers)*

A. Record
B. Group
C. Group Sort Selection
D. Record Sort Selection
E. Saved Data

Q25. Where can you view the full code generated by your report?

A. Click Show Formula – Formula Workshop
B. Click Show Formula – Formula Custom Function
C. Click Show Formula – Formula Browser
D. Click Formula Editor

Q26. **You apply the following formula within the Select Expert. Which of the following records will appear?**

({Sample_Exam_Candidate.CITY} = "Manchester" or {Sample_Exam_Candidate.CITY} = "Birmingham")
or ({@score is number} in 50 to 60)

Record No.	City	Country	Score	Exam ID
1	Birmingham	England	0	RDCR08201
2	Manchester	England	80	RDCR08301
3	Manchester	England	90	RDCR08301
4	Manchester	England	85	RDCR08301
5	Manchester	England	46	RDCR08301
6	Kent	England	70	RDCR08301
7	Essex	England	60	RDCR08301
8	Coventry	England	79	RDCR08301
9	Manchester	England	79	RDCR08301_UPGRD
10	Manchester	England	66	RDCR08301_UPGRD
11	London	England	77	RDCR08301_UPGRD
12	London	England	88	RDCR08401
13	London	England	90	RDCR08401
14	London	England	89	RDCR08401
15	London	England	90	RDCR08701
16	London	England	89	RDCR08701
17	London	England	79	RDCR08501
18	London	England	50	RDCR08501
19	London	England	100	RDCR08501
20	London	England	56	RDCR08701

A. 1,5,7,9,10,18,20

B. 7,16,5

C. 7,18,20

D. 10,4,13,15

E. 3,7,8,10

Q27. You apply the following formula to the Select Expert Record section of your report. Which of the following records will appear?

({Sample_Exam_Candidate.CITY} = "London" and {Sample_Exam. ExamID} = " RDCR08301") or ({@score is number} in 90 to 100)

Record No.	City	Country	Score	Exam ID
1	Birmingham	England	0	RDCR08201
2	London	England	80	RDCR08301
3	London	England	90	RDCR08301
4	London	England	85	RDCR08301
5	London	England	46	RDCR08301
6	Kent	England	70	RDCR08301
7	Essex	England	60	RDCR08301
8	Coventry	England	79	RDCR08301
9	London	England	79	RDCR08301_UPGRD
10	London	England	66	RDCR08301_UPGRD
11	London	England	77	RDCR08301_UPGRD
12	London	England	88	RDCR08401
13	London	England	90	RDCR08401
14	London	England	89	RDCR08401
15	London	England	90	RDCR08601
16	London	England	89	RDCR08601
17	London	England	79	RDCR08601
18	London	England	50	RDCR08601
19	London	England	100	RDCR08601
20	London	England	56	RDCR08701

A. 5,17,18,19

B. 2,6,8,19

C. 1,2,3,4

D. 2,3,4,5,13,15,19

Q28. The current date is the 23rd of June 2008; you apply the following formula within the Select Expert Record. Which of the following records will be returned?

{Orders.Order Date} in MonthToDate

Record No	Product Name	Price	Product Class	Order Date
1	MS SBS	£764.85	Server	30/04/2008
2	MS SBS	£764.85	Server	24/04/2008
3	MS SBS	£764.85	Server	13/04/2008
4	MS SBS	£764.85	Server	12/08/2007
5	MS SBS	£764.85	Server	10/08/2007
6	MS SBS	£764.85	Server	08/08/2007
7	MS SBS	£764.85	Server	26/08/2007
8	MS SBS	£764.85	Server	15/08/2007
9	MS SBS	£764.85	Server	06/06/2008
10	MS SBS	£764.85	Server	14/06/2008
11	MS SBS	£764.85	Server	03/06/2008
12	MS SBS	£764.85	Server	02/10/2007
13	MS SBS	£764.85	Server	26/10/2007
14	MS SBS	£764.85	Server	21/10/2007
15	MS SBS	£764.85	Server	28/10/2006
16	MS SBS	£764.85	Server	10/10/2005
17	MS SBS	£764.85	Server	22/10/2005
18	MS SBS	£764.85	Server	10/10/2005
19	MS SBS	£764.85	Server	13/10/2005

A. 9,10,11
B. 2,5,4, 19
C. 1,3
D. 2
E. 3,4,19

63

Q29. The current date is the 10th of April 2008; you apply the following formula within the Select Expert. Which of the following records will NOT be returned?

{Orders.Order Date} in Next30Days

SBS 2003

No	Client Name	Order No	Merch No	Unit Price	Quantity	Cost	Order Date
1	Examhints	1570	2208	£53.90	3	£161.70	25/04/2008
2	Projection Programmers	1755	2209	£53.90	3	£161.70	21/06/2008
3	Folks Books	1557	2215	£53.90	1	£53.90	22/04/2008
4	Books and Books	1769	2212	£53.90	3	£161.70	25/06/2008

SBS 2008

No.	Client Name	Order No	Merch No	Unit Price	Quantity	Cost	Order Date
5	Iroko Books	1537	1101	£14.50	3	£43.50	17/04/2008
6	Ink Only	3170	1104	£14.50	2	£29.00	02/05/2008
7	Great Stone	2985	1102	£14.50	2	£29.00	29/04/2008
8	Great Stone	1487	1104	£14.50	3	£43.50	05/03/2008
9	Iroko Books	2884	1104	£14.50	3	£43.50	03/03/2008
10	Stationary Gallore	1488	1105	£14.50	3	£43.50	05/03/2008
11	Great Stone	1666	1104	£14.50	3	£43.50	01/03/2008

A. 2,3,4,7

B. 1,3,5,6,7,

C. 1,4,6,11

D. 6,7, 11,10

Q30. Formulas can be created via the Select Expert

A. True

B. False

Q31. A Record Selection can be applied via the Standard Report Creation Wizard

A. True

B. False

Q32. The Record Selection criteria applied within the Select Expert will appear in which of the following sections of the SQL statement when viewing the Show SQL Query within Crystal Reports?

A. Select

B. Where
C. Order by
D. Group By

Q33. **Your report is grouped by Exam_id which displays RDCR08201, RDCR08301 AND RDCR08401, you have set the Select Expert - Group to One Of RDCR08301 AND RDCR08401. Which of the following statements are true?** *(Multiple Answers)*

A. The report will only display data for RDCR08301 AND RDCR08401
B. Exams RDCR08201, RDCR08301 AND RDCR08401 will still appear in the Preview Panel
C. Exams RDCR08201, RDCR08301 AND RDCR08401 will still not appear in the Preview Panel
D. but there will be no data for RDCR08201 in the report
E. Exam RDCR08401 will not still appear in the Preview Panel
F. There will be no data for RDCR08201 in the report

Q34. **A user has raised a query regarding the none retrieval of information when the Select Expert is used, the user entered field name is equal to 'iroko' in the Select Expert Record Selection of the report, the report has been refreshed however data is not being returned, you have queried the database and data for 'Iroko' exist. Which of the following could explain the problem?**

A. The user did not verify the report before refreshing it
B. The user did not verify the database before refreshing the report
C. The user should enter startwith iroko rather than equals iroko
D. The Database Is Case Sensitive, the user should go into File- Report Options and check the Database Server Is Case-Insensitive

Q35. **You have selected the following Product Types: - "Crystal Reports XI R2" and " Crystal Reports 2008"; you want to amend the Record Selection created within the Select Expert to add the " Crystal Reports XI". Which of the following methods apply?** *(Multiple Answers)*

A. From the Menu Bar select Report – Select Expert – Record - select the Product_type tab and from the drop-down menu beside Is One Of select Crystal Reports XI

B. From the Menu Bar select Report – Select Expert – Record – Formula Editor, within the Formula Editor amend the formula to include the Crystal Reports XI {Product.Product Name} in [Crystal Reports XI R2" ," Crystal Reports 2008", " Crystal Reports XI"]

C. From the Menu Bar select Select Expert – Select Expert Expert – select the Product_type tab and from the drop-down menu beside Is One Of select Crystal Reports XI

D. From the Menu Bar select Select Expert – Show Formula – Formula Editor, within the Formula Editor amend the formula to include the Crystal Reports XI : {Product.Product Name} in [Crystal Reports XI R2" ," Crystal Reports 2008", " Crystal Reports XI"]

Q36. **You select a number field within the Select Expert and leave the 'Is Equal' section Blank. Which of the following will occur?**

A. {TableName.FieldName} = ""
B. An error message will be produced
C. The OK button will be greyed out
D. {TableName.FieldName} ""

Q37. **You have saved data with your report, you do not want to query the database directly, however you want to apply the record selection based on the saved data. Which of the following apply?**

A. Report – Select Expert - Record
B. Report – Select Expert - Group
C. Report – Select Expert – Saved Data
D. None of the above, this functionality does not exist

Q38. **When a new report is created and refreshed it queries the database directly**

A. True
B. False

Q39. You are refreshing a report after you have changed the record selection, which of the following refresh options will be available? *(Multiple Answers)*

A. Use Saved data
B. Refresh Report
C. Refresh Data
D. Saved Data only

Q40. Is in the Period and Is not in the Period are only applicable to date fields

A. True
B. False

Q41. Which of the following procedures needs to be implemented to activate interactive sorting via the viewer? *(Multiple Answers)*

A. Add database fields to the Record Sort Expert
B. Remove database fields to the Record Sort Expert
C. Bind database fields to existing sorted fields
D. Unbind database fields from existing sorted fields

Q42 You want to show a new page after every four customers in your group. Which of the following apply? *(Multiple Answers)*

A. Select Report - Group Expert – Highlight the Customer Name under group by, select options – Options Tab – Click New Page After and enter 4 under Visible Groups
B. Select Report – Record Sort Expert – Highlight the Customer Name under group by, select options – Options Tab – Click New Page After and enter 4 under Visible Groups
C. Right – click the section – Change Group – Options Tab – Click New Page After and enter 4 under Visible Groups
D. Select Report – Section Expert – Highlight the Customer Name under group by, select options – Options Tab – Click New Page After and enter 4 under Visible Groups

Q43 You have applied a record selection of Customer City = London within the Select Expert – Record; results returned equals 100 customers, you are not sure these records are correct. (The backend database is SQL Server) How can you verify the data retrieved? *(Multiple Answers)*

A. Within SQL Server type in the required code and verify record.
B. Within Crystal Reports – select Database – Show SQL Query – copy the query into SQL and verify
C. Within Crystal Reports – select Report – Show SQL Query – copy the query into SQL and verify
D. Within Crystal Reports – select File – Show SQL Query – copy the query into SQL and verify

Q44. Which methods can be used to limit the records retrieved? *(Multiple Answers)*

A. Select Expert Record
B. Select Expert Group
C. Select Expert Saved Data

Organize Data in a Report

Q1. **You want to customize the group within your report. Which of the following apply?**

A. Right-click the group and select Change Group, from the drop-down menu select in Specified Order, in the Name Group section enter the name of your first specified group and click new, from the Define Named Group dialog box, select Is One Of from the dropdown list and add the data applicable to the first specified group and click ok, click the Others tab and select Put All Others Together With The Name you specify and click ok

B. Right-click the group and select Section Expert, from the drop-down menu select in Specified Order, in the Name Group section, enter the name of your first specified group and click new, from the Define Named Group dialog box, select Is One Of from the dropdown list and add the data applicable to the first specified group and click ok, click the Others tab and select Put All Others Together With The Name you specify and click ok

C. Right-click the group and select Group Section from the drop-down menu select in Specified Order, in the Name Group section, enter the name of your first specified group and click new, from the Define Named Group dialog box, select Is One Of from the dropdown list and add the data applicable to the first specified group and click ok, click the Others tab and select Put All Others Together With The Name you specify and click ok

Q2. **You have applied a specified order to your group, this will not be reflected in the Group section of the Preview Panel.**

A. False
B. True

Q3. **You want to change the date format of your group from each day to each month. Which of the following apply?**

A. Right-click the Order Date group and select Section Expert, under 'The Section Will Be Printed' select for each month
B. Right-click the Order Date group and select group section, under 'The Section Will Be Printed' select for each month

C. Right-click the Order Date group and select group, under 'The Section Will Be Printed' select for each month
D. Right-click the Order Date group and select Change Group, under 'The Section Will Be Printed' select For Each Month

Q4. **Which of the following are options available under the Change Group section, when applying a change to a date group?** *(Multiple Answers)*

A. For each day, For each quarter
B. For each Month, For each half year
C. For each record
D. For each two weeks
E. For each half month
F. For AM/PM,
G. For each MM\DD\YYYY
H. For each Week
I. For each Hour

Q5. **You want to prevent your group from breaking over several pages. What should you do?**

A. Right-click the grey section of the group, select Change Group, select the Options Tab, and check the Keep Group Together checkbox
B. Select database, check the Keep Group Together checkbox
C. Select report, check the Keep Group Together checkbox
D. Select Format, check the Keep Group Together checkbox

Q6. **Your date group spills over more than one page, however the heading for the group does not appear on the second page. What can you do to ensure the group title appears on each page?**

A. Select database, check the Keep Group Together checkbox
B. Select report, check the Keep Group Together checkbox
C. Right-click the grey section of the group, select Change Group, select the Options Tab, and check the Repeat Group Header On Each Page
D. Select Format, check the Repeat Group Header On Each Page

Q7. **Which of the following statements are true?** *(Multiple Answers)*

A. The Group Sort Expert will be grayed out unless a group summary has already been created on the report.
B. The Group Tree will appear irrespective of the creation of a group
C. The Group Sort Expert will work, without a group
D. Grouping your report enables the Group Tree in the Preview Panel

Q8. **You want to create a Dynamic Group based on options 1 to 3 below. Which of the following apply?**

1. Exam City
2. Exam Center
3. Exam Score

A. if GroupName ({Exam.Exam Center}) = "Exam Center" then {Exam. Center } else if GroupName ({Exam.Exam City}) = "City" then {Exam. City } else if GroupName ({Exam. Exam Score}) = Exam Score then {Exam.Exam Score}
B. Create a parameter (Group Type) and enter options 1 to 3 in the pick list, create a formula as follows: if {? Group Type} = "Exam Center" then {Exam. Center } else if {? Group Type} = "City" then {Exam.City } else if {? Group Type} = "Exam Score" then {Exam. Score }, insert the formula as the group, Insert a candidate summary count based on the group
C. Create a parameter (Group Type) and enter options 1 to 3 in the picklist, insert the parameter as the group within the report, then create a formula as follows: if {? Group Type} = "Exam Center" then {Exam. Center } else if {? Group Type} = "City" then {Exam. City } else if {? Group Type} = "Exam Score" then {Exam. Score }, Insert a summary count based on the group
D. This cannot be done, dynamic groups do not work within Crystal Reports

Q9: **You can drill down from your chart into the details section. Which chart layout has been used?**

A. Group
B. Report
C. OLAP

D. Header

E. Advanced

Q10: You want to add a group to your report. Which of the following can achieve this?

A. Choose Insert | Group | select group field and click ok

B. Choose Format | Group | select group field and click ok

C. Choose Report | Group Expert | select group field and click arrow and ok

D. Choose Edit | Group | select group field and click ok

Q11. You have inserted a section below the details section, the section name will be renamed as which of the following?

A. Details i, Details ii

B. Details a, Details b

C. Details Aa, Details Bb

Q12. Using the following Sales information in the diagram below; you have created a summary of product sales for the month of July, and would like to display the top three product sales for this month. Which of the following apply?

Product Name	Sales for Month July
Crystal Report 2008	£5000.00
Crystal Report 8	£90.00
Crystal Report XIR2	£3824.25
Crystal Report XIR1	£2159.4
Crystal Report 2008	£6000
Crystal Report 9	£203.4
Crystal Report 2008	£7500
Crystal Report 2008	£548.7
Crystal Report 7	£72.5
Crystal Report 7	£99
Crystal Report 2008	£3298.5
Crystal Report 2008	£8000
Crystal Report 8	£35.4

D. Select Format – Group Sort Expert – for this Group Sort select 'TopN' based on the sales summary for July and where N is 3 and click OK

E. Select Insert – Group Sort Expert – for this Group Sort select 'TopN' based on the sales summary for July and where N is 3 and click OK

F. Select Report –Record Sort Expert – for this Group Sort select 'TopN' based on the sales summary for July and where N is 3 and click OK

G. Select Report – Group Sort Expert – for this Group Sort select 'TopN' based on the sales summary for July and where N is 3 and click OK

Q13. **You have created a summary of product sales for the Month of July which displays the top three product sales for this month. You want all other sales to be grouped under the title 'Other Sales'. Which of the following apply?**

A. Select Format – Group Sort Expert –and check the include others with the name 'Others Sales for this period' and click OK

B. Select Report – Group Sort Expert – and check the include others with the name 'Others Sales for this period' and click OK

C. Select Insert – Group Sort Expert – and check the include others with the name 'Others Sales for this period' and click OK

D. Select Report –Record Sort Expert –and check the include others with the name 'Others Sales for this period' and click OK

Q14. **By checking the 'include ties' in the Group Sort Expert. Which of the following will happen?**

A. It will include individual fields whose summarized values are equal.
B. It will include Report Headers whose summarized values are equal.
C. It will include groups whose summarized values are equal.
D. It will include individual formulas whose summarized values are equal.

Q15. **Which of the following statements apply to the addition of a summary to your report?** *(Multiple Answers)*

A. Choose Insert | Summary | Choose field and type of calculation (sum, average, maximum, minimum) select the summary location, by group or by grand total and click OK

B. Insert Group Summary | Choose type of calculation (sum, average, maximum, minimum) select the summary location, by group or by grand total and click OK

C. Insert Group | Choose type of calculation (sum, average, maximum, minimum) select the summary location, by group or by grand total and click OK

D. Right-click field to summarize | Choose field and type of calculation (sum, average, maximum, minimum) select insert summary, select the summary location, by group or by grand total and click OK

Q16. You want to prevent groups from spilling over to other pages. What can you do to ensure this does not happen?

A. Use Repeating Headers
B. Use RepeatingFooters
C. Use RepeatedGroupHeaders
D. Use Keep Together

Q17. Within the Group Sort Expert which of the following are not available? *(Multiple Answers)*

A. No sort
B. Bottom totals
C. All
D. Top N
E. Bottom N
F. Top Percentage
G. Bottom Percentage
H. Top totals N

Q18. You have created an Ascending Sort Order on a field which contains both numeric and text data. Which one is true?

A. The string fields will appear before the numbers field
B. You cannot perform a sort on a numeric field
C. You can perform a sort on the Text Object
D. The numbers will appear before the string fields

Q19. You add a group to a report. This action will produce a Group Header and a Group Footer.

A. True
B. False

Q20. Which of the following cannot be implemented through the Change Group Options? *(Multiple Answers)*

A. Ascending or Descending
B. Specified
C. Original
D. Keep Group Together
E. Repeat Group Header On Each New Page
F. Format Group Expert
G. Sum Group

Q21. When a group is inserted into a report a Group Header and Group Footer are created.

A. True
B. False

Q22. You have excluded sales which do not fall into the Top 3 sales category, however they still appear in the grand total. Which of the following methods will resolve this?

With others included

Product Name	Sales for Month August
Crystal Report 2008	£8000
Crystal Report XI R2	£7500
Crystal Report XI	£6000
Other Sales for this Period	£15,331.15
	£36,831.15

With others excluded

Product Name	Sales for Month August
Crystal Report 2008	£8000
Crystal Report XI R2	£7500
Crystal Report XI	£6000
	£36,831.15

A. Verify the database and the Grand Total will change to reflect the TopN sales figure
B. Create a running total based on the sales amount and place it in the report footer
C. Select View and Field Explorer - Right-click Running Total and select New, enter running total name, and summarize sales amount with a summary type of sum and click ok
D. None of the above

Q23. Summaries based on the Group Footer can be used in the alert creation process.

A. True
B. False

Q24. You have inserted a group into your sales report which you would now like to delete. What should you do?

A. Choose Format | Delete Group
B. Right-click the grey section of the group and select delete group from the drop-down menu
C. Choose Report | Delete Group
D. From the Section Expert, Delete Group
E. Select Edit and delete group

Q25. You have created a report with a country group, and you want users to have the functionality to drill down from a Country of their choice to view the customers associated with that Country. What should you do? *(Multiple Answers)*

A. Right-click the grey area of the detail section and select Hide (Drill-Down OK), when you double-click the group the Details Section will appear.
B. From the toolbar, select Report | Section Expert | highlight the Details Section and check the hide ((Drill-Down OK), when the user double-clicks the group the Details Section will appear.
C. Right-click the grey area of the Details Section and select suppress, when you double-click the group the Details Section will appear.
D. Right-click the grey area of the detail section and select do not show section, when you double-click the group the Details Section will appear.

E. Right-click the grey area of the Group and select do not show section, when you double-click the group the Details Section will appear.

F. Right-click the grey area of the detail section and select hide ((Drill-Down OK) from the drop-down list

Q26. You have created a Finance report and would like to see the 'Customer. Name' records appearing in an ascending order. What should you do?

A. Choose Report | Select Expert from the pull-down menu, or click the record sort expert button from the toolbar, from the available fields section, select Customer name field and use the arrow > to place it into the sort-fields on your right, click the radio button under sort direction   (Ascending)

B. Choose Report | Section Expert | Record Sort Expert from the pull-down menu, or click the record sort expert button from the toolbar, from the available fields section, select Customer name field and use the arrow > to place it into the sort-fields on your right, click the radio button under sort direction   (Ascending)

C. Choose Report | Record Sort Expert | from the pull-down menu, or click the record sort expert button from the toolbar, from the available fields section, select {Customer. name} field and use the arrow > to place it into the sort-fields on your right, click the radio button under sort direction   (Ascending)

D. Right-click field, select format field | Common Tab, Ascending

Q27. You want to create a summary report based on the number of clients per country. Which of the following will enable you to create the summary report?

A. Select file new and Blank Report , select the data source required and select the client table, click finish, from the insert group dialog box select Client_id as the group and click OK, drag the client_id field from the Field Explorer into the Details Section of the report and right-click the field and insert summary, select distinct count under calculate this summary and for the summary location, select the Country group and click OK, hide the detail section and preview

B. Select file new and Blank Report from the Menu Bar, the Database

Expert will appear, select the data source required and select the client table, click finish, from the detail section select insert from the Menu Bar a select group, from the insert group dialog box select country as the group and click OK, drag the client_id field from the Field Explorer into the Details Section of the report and right-click the field and insert summary, under calculate this summary select distinct count and for the summary location, select the Country group and click OK, hide the detail section and preview

C. Select file new and Blank Report from the Menu Bar, the Database Expert will appear, select the data source required and select the client table, click finish, from the detail section select insert from the Menu Bar a select group, from the insert group dialog box select country as the group, drag the client_id field from the Field Explorer into the Details Section of the report and right-click the field and insert summary, select sum under calculate this summary and for the summary location, select the Country group and click OK, hide the detail section and preview

D. Select file new and Blank Report from the Menu Bar, the Database Expert will appear, select the data source required and select the client table, click finish, and click OK, drag the client_id field from the Field Explorer into the Details Section of the report and right-click the field and insert summary, select distinct count under calculate this summary and for the summary location, and click OK, Suppress the detail section and preview

Q28.   You have created a report with one group, and you do no want the Details Section to appear when the user double clicks the group. What should you apply to the details section?

A. Choose Report from the Menu Bar | select selection expert | highlight the Details Section and check Suppress

B. Choose Report from the Menu Bar | select Section Expert | highlight the Details Section and check the Hide (Drill-Down OK) checkbox

C. Choose Report from the Menu Bar | select Section Expert | highlight the Details Section and check Suppress

D. Choose Report from the Menu Bar | select Section Expert | highlight the Details Section and check Keep Together

Q29. **You want to apply a Group Selection criteria to your report. What should you do?**

A. From the menu bar select Report - Select Expert - Group – select the field – Formula Editor - and enter your criteria

B. From the Select Expert, click the record selection radio button, select formula editor and enter your criteria

C. From the Select Expert, click the records tab, select formula editor and enter your criteria

D. From the select Section Expert, click the Group Selection radio button, select formula editor and enter your criteria

Q30. **You have created a formula which outlines early and late deliveries. You have also created a group based on this formula, and want to highlight all Group Header backgrounds in red, if delivery is late. Which of the following apply?**

A. Right-click grey section of group and select - Select Expert, select the color tab and click the background color formula box and enter the following formula if {@DeliveryStatus} = 'Late Delivery' then Crred else CrNoColor

B. Right-click grey section of group and select Section Expert, select the Common Tab and click the background color formula box and enter the following formula if {@DeliveryStatus} = 'Late Delivery' then Crred else CrNoColor

C. Right-click grey section of group and select Section Expert, select the color tab and click the background color formula box and enter the following formula if {@DeliveryStatus} = 'Late Delivery' then Crred else CrNoColor

D. Right-click grey section of group and select Section Expert, select the Font tab and click the background color formula box and enter the following formula if {@DeliveryStatus} = 'Late Delivery' then Crred else CrNoColor

Q31. **You want to establish the percentage of orders by clients for the month of July. Which of the following apply?** *(Multiple Answers)*

A. Create the following formula and place it in the client Group

Header Sum ({Merchandise.Cost}) % Sum ({Merchandise.Cost}, { Merchandise. Merchandise Name})

B. Sum ({Merchandise.Cost}, { Merchandise. Merchandise Name}) %

C. Check the Show As Percentage of checkbox when inserting a summary

D. Create the following formula and place it in the client Group Header Sum ({Merchandise.Cost}, { Merchandise. Merchandise Name}) % Sum ({Merchandise.Cost})

Q32. **You have drilled-down several levels within a report. Which of the following are true?** *(Multiple Answers)*

A. Only the material appearing in the current tab will export
B. You can print all material in the report after drill-down
C. You cannot export drill-down information
D. Only the material appearing in the current tab will print

Q33. **A request had been made for a report, the report should be group specific, the report must display only departments with over 15 staff; the report contains a department field and a distinct count of department staff and is grouped by department. Which of the following will achieve this result?**

A. Create a distinct count of staff based on the department group, from the toolbar select Report and Select Expert and click the Show Formula button, and also check the Group Selection Button, and click on the Formula Editor, this will open up the Formula Workshop - Group Selection Formula Editor, enter the formula below based on the group summary created, DistinctCount ({Department.Staff}, { Department.Department}) > 15, save and close and click OK

B. Create a distinct count of staff based on the department group, from the toolbar select Report and Select Expert and click the Show Formula button, and also check the Group Selection Button, and click on the Formula Editor, this will open up the Formula Workshop - Group Selection Formula Editor, enter the formula below based on the group summary created, Count ({Department.Staff}, {Department.Department}) > 15, save and close and click OK

Q34. You have created a report which contains a date group, the date field used to create the group contains a Null field, you created a formula to display the word 'Blank' instead of displaying a Null field in the Group Tree, the word 'Blank' is not appearing in the Group Tree, what could the problem be?

A. Convert Database NULL Values to Default is checked
B. Convert Database NULL Values to Default is unchecked

Q35. Grouped data spills over onto the next page of your report. You want to highlight this information with the group title: Group Name (Continued from previous page). Which of the following will achieve this? *(Multiple Answers)*

A. Create the following formula and place it in Group Header #2: if InRepeatedGroupHeader = true then {Sales_3.Region} + " (.......
Continued from previous page)"
B. if InRepeatedGroupHeader = False then {Sales_3.Region} + "
(.......Continued from previous page)"
C. Create the following formula and place it in Group Header #2:
if InRepeatedGroupHeader = yes then {Sales_3.Region} + " (.......
Continued from previous page)"
D. Create the following formula and place it in Group Header #2:
if InRepeatedGroupHeader = no then {Sales_3.Region} + " (.......
Continued from previous page)"

Q36. You have created a report with two group summaries, Group Header #1 is based on Continent Sales and Group Header #2 which the user will drill-down into is based on Regional Sales, when the user drills-down from continent sales into regional sales, Group Header #1 still appears at the top, you only want to see the sales information for regional sales. How can you implement this? *(Multiple Answers)*

A. Within the Section Expert apply the following conditional formula by clicking the suppress conditional button for Group Header #2 (Region Group) and entering the following condition DrillDownGroupLevel = 1
B. Within the Section Expert apply the following conditional formula by clicking the suppress conditional button for Group Header

#1 (Continent Group) and entering the following condition DrillDownGroupLevel <> 0

C. Within the Section Expert apply the following conditional formula by clicking the suppress conditional button for Group Header #2 (Region Group) and entering the following condition DrillDownGroupLevel <1

D. Within the Section Expert apply the following conditional formula by clicking the suppress conditional button for Group Header #1 (Continent Group) and entering the following condition DrillDownGroupLevel = 1

Q37. You want to view the restriction based on the groups within your report. Which of the following apply?

A. Create a formula within your report and enter GroupSelection
B. Create a formula within your report and enter GroupSelect
C. Create a formula within your report and enter RecordSelection
D. Create a formula within your report and enter RecordSelect

Q38. When you suppress the group header the group footer will also be suppressed

A. True
B. False

CHAPTER 2 - CUSTOMIZE AND FORMAT A REPORT

Chapter 2 covers questions on formatting a report, adding objects and customised objects, it also covers adding graphical elements, applying section formatting and formatting data conditionally, creating charts and chart types and apply report templates.

Keywords:
Charts, Group Charts, Advanced Charts Format, Objects, Text Interpretations, Highlight, Lock Position, Watermark, Dynamic Location, Multiple Columns, Default, Template Field Object

Format Objects and Format Data Conditionally

Q1. **You want to highlight (in Red) all sales that have fallen below or equal the 1000 minimum sales target set for each month. Your report is grouped by the order date and you have created a sum of total sales based on the order date. Which of the following apply?** *(Multiple Answers)*

A. From the Menu Bar select format – Format Field – select the font tab, click the formula box X+2 beside calculation and enter: if {Sales.SalesTotal} <= 1000 then Crred

B. From the Menu Bar select format – Format Field – select the font tab, click the formula box X+2 beside color and enter: if {Sales.SalesTotal} > 1000 then crrred

C. Select the Sum of Total Sales, from the Menu Bar select format – Format Field – within the Format Editor select the font tab, click the formula box X+2 beside color and enter: if {Sales.SalesTotal} < 1000 then crrred

D. From the Menu Bar select format – Format Field – select the font tab, click the formula box X+2 beside color and enter: if {Sales.SalesTotal} < 1000 then red

E. Right-click the SalesTotal field and select Format Field – within the Format Editor select the font tab, click the formula box X+2 beside color and enter: if {Sales.SalesTotal} <= 1000 then Crred

Q2. Which of the following are Date orders in the Format Editor? *(Multiple Answers)*

A. DDY
B. DMY
C. MDY
D. YMD

Q3. Which of the following has a floating currency applied and which one has a Fixed Currency option applied?

A.

£ 43.50
£ 67.80
£ 5,237.55
£ 2,792.86

£43.50
£67.80
£5,237.55

B.
£2,792.86

Q4. Which of the following Text Interpretations types are available via the Format Editor? *(Multiple Answers)*

A. XML Text
B. HTML Text
C. RTF Text
D. TXT Text

Q5. You want to apply AM and PM to your time field, under which tab in the Format Editor would you find this function

A. Time
B. Date
C. Date and Time
D. Date only

Q6. **You notice the AM and PM functions are greyed out. Which of the following could be the cause?**

A. The 24 hour button is selected
B. The 12 hour button is selected
C. Use System Default Format is checked
D. Symbol position is not set

Q7. **The comments field contains long text, you want to display all text on your report without compromising the number of fields you already have in your report. How can you ensure that all text in the comments field are displayed?**

A. Tick the Can Grow checkbox from the Font Tab of the Format Editor
B. Tick the Can Grow checkbox from the Common Tab of the Format Editor
C. Tick the Can Grow checkbox from the Border Tab of the Format Editor
D. Tick the Can Grow checkbox from the Paragraph Tab of the Format Editor

Q8. **You want to highlight all Employees who are employed in New York; the field must be displayed in Tahoma. Which of the following apply?**

A. Select the City Field, from the Menu Bar select format – Format Field – select the font type tab, click the formula box X+2 beside calculation and enter: if {Employee.City} = 'New York' then "Tahoma"
B. Select the City Field, from the Menu Bar select format – Format Field – select the font tab, click the formula box X+2 beside font and enter: if {Employee.City} = 'New York' then "Tahoma"
C. Select the City Field, from the Menu Bar select format – Format Field – select the color tab, click the formula box X+2 beside font and enter: if {Employee.City} = 'New York' then "Tahoma"
D. Select the City Field, from the Menu Bar select format – Format Field – select the font tab, click the formula box X+2 beside font and enter: if {Employee.City} is like 'Lon' then "Tahoma"

Q9. **The text in your report appears as follows:**

700 South Kingsway
Which setting will change it back to?
South Kingsway 700

    A. Right-click the field and select Format Field, click the Paragraph Tab, under Reading order, set back to Right to Left from Left to Right
    B. Right-click the field and select Format Field, click the Paragraph Tab, under Reading order, set back to Left to Right from Right to Left
    C. Right-click the field and select Format Field, click the font tab, under Reading order, set back to Left to Right from Right to Left
    D. Right-click the field and select Format Field, click the Common Tab, under Reading order, set back to Right to Left from Left to Right

10. **Your text appears as follows: Which of the following has been applied to the text?**

> South Kingsway 7464
> Eighth Avenue 410

    A. Border Shadow
    B. Tight Horizontal
    C. Drop Shadow
    D. Border background color

Q11. **You want the currency symbol ($) to appear after the currency. Where can you implement this setting within the Format Editor?**

    A. Right-click field – Format Field – Format Editor -Number Tab – Customized - Currency Symbol Tab – check the Enable Currency Symbol and select the position required
    B. Right-click field – Format Field – Format Editor -Currency Number Tab - Currency Symbol Tab and Position
    C. Right-click field – Format Field – Format Editor -Currency Symbol Tab - Number Tab and Position
    D. Right-click field – Format Field – Format Editor -Position Tab - Number Tab and Position

Q12. **You want to provide users with the functionality to email sales representatives directly within your report. You have email addresses stored in the database table, which you have placed on the report. Which of the following methods apply?**

A. Right-click the report and select Format Field, click the Format tab and select Current E-mail Field Value from the Hyperlink Type
B. Right-click the report and select Format Field, click the Report tab and select Current E-mail Field Value from the Hyperlink Type
C. Right-click the field and select Format Field, click the Hyperlink Tab and select Current E-mail Field Value as the Hyperlink Type
D. Right-click the report and select Format Field, click the Paragraph Tab and select Current E-mail Field Value from the Hyperlink Type

Q13. **You want to apply a color to a field's text. Where can this be applied within the Format Editor?**

A. Border Tab
B. Common Tab
C. Paragraph Tab
D. Hyperlink Tab
E. Font Tab

Q14. **You want to apply conditional formatting to a report field. Which of the following can you use?** *(Multiple Answers)*

A. Format Editor
B. Selection Expert
C. Group Expert
D. Group Section
E. Field Editor
F. Highlighting Expert

Q15. **How can you change a date format to ddmmyyyy?**

A. Right-click the field, select Format Field, click the Date and Time tab, click Customize, click the Date Tab, select the DMY button under order and in the format section set Month to 3, Day to 1 and Year to 1999 and click ok

B. Right-click the field, select Format Field, click the Date and Time tab, click Customize, click the Date Tab, select the DMY button under order and in the format section change Month to 03, Day to 01 and Year to 1999 and click ok

C. Right-click the field, select Format Field, click the Date and Time tab, click Customize, click the Date Tab, select the DMY button under order and in the format section change Month to 03, Day to 1 and Year to 1999 and click ok

D. Right-click the field, select Format Field, click the Date and Time tab, click Customize, click the Date Tab, select the DMY button under order and in the format section change Month to 03, Day to 01 and Year to 99 and click ok

Q16. You do not want users to move fields within the report. Which of the following apply?

A. Tick the Lock Position and Size checkbox from the Paragraph Tab of the Format Editor

B. Tick the Lock Position and Size checkbox from the Common Tab of the Format Editor

C. Tick the Lock Position and Size checkbox from the Border Tab of the Format Editor

D. Tick the Lock Position and Size checkbox from the Font Tab of the Format Editor

E. Tick the Read-Only Checkbox from the Common Tab of the Format Editor

Q17. Your Boolean formula displays a True or False result. You want to change it to Y or N. Which of the following apply?

A. Right-click the field, select Format Field, click the Formula tab and select Y or N from the Boolean Text drop down menu

B. Right-click the field, select Format Field, click the String tab and select Y or N from the Boolean Text drop down menu

C. Right-click the field, select Format Field, click the Common Tab and select Y or N from the Boolean Text drop down menu

D. Right-click the field, select Format Field, click the Boolean tab and select Y or N from the Boolean Text drop down menu

Q18. **You want to stop objects being formatted. Which of the following apply?** *(Multiple Answers)*

A.   Right-click object and select Format Field, from the drop-down menu, select the Common Tab and place a tick in the read only check box

B.   Highlight field, from the Menu Bar select format – Format Field, select the Common Tab and place a tick in the Lock Position and Size check box

C.   Select object, right-click and select No format

D.   Select object, right-click and select Format, Field type, Read Only

E.   Highlight field, from the Menu Bar select format – Format Field, select the Common Tab and place a tick in the read only check box

Q19. **You want to apply a 270-degree rotation to your text. What should you do?**

A.   Right-click the field, select Format Field, click the Common Tab, and under Text rotation select 90 degrees from the drop-down list.

B.   Right-click the text, select Format Field, and click the Paragraph Tab, under Text rotation 270 degrees from the drop-down list.

C.   Right-click the text, select Format Field, and click the format tab, under Text rotation and degrees 270 degrees from the drop-down list.

Q20. **How would you suppress duplicate records?**

A.   Right-click the field, select Format Field, click the Paragraph Tab, and check the suppress If Duplicated checkbox

B.   Right-click the field, select Format Field, click the Font tab, and check the suppress If Duplicated checkbox

C.   Right-click the field, select Format Field, click the Common Tab, and check the suppress checkbox

D.   Right-click the field, select Format Field, click the Common Tab, and check the Suppress Duplicated records only checkbox

E.   Right-click the field, select Format Field, click the Common Tab, and check the Suppress If Duplicated checkbox

Q21. You want to create a borderline at the bottom of your text. Which of the following apply?

A. Right-click the field, select Format Field, click the Paragraph Tab and apply a single line from the drop-down menu to the Bottom section of the field
B. Right-click the field, select Format Field, click the font tab and apply a single line from the drop-down menu to the Bottom section of the field
C. Right-click the field, select Format Field, click the Border Tab and apply a single line from the drop-down menu to the Bottom section of the field
D. Right-click the field, select Format Field, click the Common Tab and apply a single line from the drop-down menu to the Bottom section of the field

Q22. When a field is placed in the details section it appears as follows:

CR6321
CR6592
CR6798
You notice the section around the text extends as far as the longest text in the field. You want to
limit this extension to individual text length
CR6321
CR6592
CR6798

Which of the following should you apply?

A. Right-click the field, select Format Field, click the Paragraph Tab and check the Tight Vertical checkbox
B. Right-click the field, select Format Field, click the Border Tab and check the Tight Horizontal checkbox
C. Right-click the field, select Format Field, click the Paragraph Tab and check the Border Control checkbox
D. None of the above, this cannot be done

Q23. **Your dates appear as follows:**

15\06\2009
16\06\2009
17\06\2009
18\06\2009

You want to change the format to appear as below. Which of the following apply?

15-06-2009
16-06-2009
17-06-2009
18-06-2009

    A. Right-click the field, select Format Field, click the Date and Time tab, click Customize, click the Date Tab, under separators, set first and second to -
    B. Right-click the field, select Format Field, click the Date and Time tab, click Customize, click the Date Tab, under separators, set first and second to -
    C. Right-click the field, select Format Field, click the Date and Time tab, click Customize, click the Time Tab, under separators, set first and second to -
    D. Right-click the field, select Format Field, click the Date and Time tab, click Customize, click the Date Tab, under separators, set first and second to --

Q24. **You want to apply a Tool Tip to a field. Which of the following applies?**

    A. Right-click the field, select Format Field, click the Paragraph Tab and enter the text in the Tool Tip Text space
    B. Right-click the field, select Format Field, click the Border Tab and enter the text in the Tool Tip Text space
    C. Right-click the field, select Format Field, click the string tab and enter the text in the Tool Tip Text space
    D. Right-click the field, select Format Field, click the Common Tab and enter the text in the Tool Tip Text space

Q25. Which tab within the Format Editor allows you to set indentations, spacing, reading order and Text Interpretation?

A. Border Tab
B. Common Tab
C. Paragraph Tab
D. Hyperlink Tab
E. Font Tab

Q26. Which of the following are line spacing types? *(Multiple Answers)*

A. Normal
B. Exact
C. Multiple
D. Single

Q27. How would you apply a currency symbol of $ to a currency field?

A. Right-click the field, select Format Field, click the Currency tab, and check the display currency symbol, click customize and select currency symbol tab, enter your symbol in the currency symbol section provided
B. Right-click the field, select Format Field, click the Monetary tab, and check the display currency symbol, click customize and select currency symbol tab, enter your symbol in the currency symbol section provided
C. Right-click the field, select Format Field, click the Finance tab, and check the display currency symbol, click customize and select currency symbol tab, enter your symbol in the currency symbol section provided
D. Right-click the field, select Format Field, click the Number tab, and check the display currency symbol, click customize and select currency symbol tab, enter your symbol in the currency symbol section

Q28. **The text in the Details Section of your report appears as follows. What action can be taken to remove the line running through the text?**

RDCR08201
RDCR08301
RDCR08401
RDCR501

A. Right-click the field, select Format Field, click the font tab, under the Effects section uncheck the underline checkbox
B. Right-click the field, select Format Field, click the font tab, under the Effects section uncheck the strikeout checkbox
C. Right-click the field, select Format Field, click the font tab, under the Effects section uncheck the line checkbox
D. Right-click the field, select Format Field, click the font tab, under the Effects section uncheck the cross-out checkbox

Q29. **Which of the following are formats available within the Hyperlink tab of the Format Editor?** *(Multiple Answers)*

A. A Website On The Internet
B. Current Website Field Value
C. An Email Address
D. A File
E. Current E-Mail Field Value

Q30. **Which of the following are DHTML Viewer Only options?** *(Multiple Answers)*

A. You can create a Hyperlink to Another Report Object
B. Report Part Drill-Down
C. Email indentation
D. Map Email Hyperlink

Q31. **Which of the following are Border Line style options available via the Format Editor?** *(Multiple Answers)*

A. None
B. Single
C. Double

D. Dashed
E. Dotted

Q32. What is the difference between Lock Position and Size, and the Read-only formats, within the Format Editor?

A. With the Lock Position and Size, you cannot move or resize the field it is applied to but you can format the field, with the Read-only format applied you cannot format or resize the field but it can be moved

B. With the Lock Position and Size, you can move, format the field or resize the field it is applied to but you cannot delete it, with the Read-only format applied you can only format or resize the field but it cannot be moved

C. With the Lock Position and Read-only format applied you cannot delete the field

D. With the Lock Position and Read-only format applied you cannot move the field

Q33. Which of the following three are the indentation types? *(Multiple Answers)*

A. First Line
B. Second Line
C. Right
D. Left
E. Top
F. Bottom

Q34. Values appear as follows within your report. Which of the following apply? *(Multiple Answers)*

$ #####
$ 30.00
$ ######
$ 80.00
$ 60.00

A. Expand Field
B. Allow Field Clipping is disabled

C. Allow Field Clipping is enabled

D. Check the Can Grow checkbox

Q35. **Which of the following Tabs are available when formatting a section via the Section Expert?** *(Multiple Answers)*

A. Common Tab

B. Border Tab

C. Paging Tab

D. Hyperlink Tab

E. Color Tab

Q36. **What is the default design layout of a Blank Report?**

A. Report Header|Page Header|Detail|Report Footer|Page Footer

B. Page Header| Report Header |Detail|Report Footer|Page Footer

C. Page Header| Report Header |Detail|Page Footer| Report Footer

D. Page Header| Detail| Report Header |Page Footer| Report Footer

Q37. **You want the detailed section to appear after a drill-down. What should you apply to the section?**

A. Choose Report from the Menu Bar | select Section Expert | highlight the Details Section and check Free-From placement

B. Choose Report from the Menu Bar | select Section Expert | within the Common Tab highlight the Details Section and check the Hide (Drill-Down OK) checkbox

C. Choose Report from the Menu Bar | select Section Expert | highlight the Details Section and check Suppress

D. Choose Report from the Menu Bar | select Section Expert | highlight the Details Section and check Keep Together

Q38. **You want to resize the Details Section of your report to remove all spaces at the bottom. What is the next step?**

A. Select arrange lines

B. Size section

C. Fit section

D. Insert line

Q39. You have created a report with several groups; you want details within each group to stay within its own grouping. Which of the following should be applied to the section?

A. New Page Before
B. Keep Together
C. Suppress Blank Sections
D. New Page After
E. Keep Group Together

Q40. You want to format a section of your report. What should you use?

A. Format Editor
B. Selection Expert
C. Group Expert
D. Group Section
E. Section Expert

Q41. You do not want users to drill-down into the details section when viewing group details. What should you apply to the details section?

A. Free-From placement should be applied to the detail section
B. Hide should be applied to the details section
C. Suppress the details section
D. Keep Together should be applied to the details section

Q42. Your report contains groups which start on new pages; you would like the page number to start from 1 upon the beginning of every new group. What should you do?

A. From the Section Expert check the New Page After
B. From the Section Expert check the New Page Before
C. From the Section Expert check the Keep Together
D. From the Section Expert check the Print at bottom of page
E. From the Section Expert – Paging Tab - check the Reset Page No After checkbox

Q43. You have applied conditional formatting to the suppress formula within the Section Expert. What will the color of the x+2 button turn to?

A. sx+2 will be green
B. x+2 will be blue
C. x+2 will be red
D. x+2 will not change

Q44. You want to insert a second details section. Which of the following apply? *(Multiple Answers)*

A. Right click the grey area of the Details Section and select insert section below
B. Right click the grey area of the Details Section and select format section and insert, delete, arrange lines and fit section.
C. Right click the grey area of the Details Section and you can select alignment, select insert, delete, arrange lines and fit section.
D. Right click the grey area of the Details Section and you can select Report format, select insert, delete, arrange lines and fit section.
E. Within the Section Expert – highlight the details section and click insert

Q45. Each time you view your report there are several blank sections. What should you do?

A. Section Expert – Common Tab - Suppress Blank Section
B. Section Expert – Paging Tab - New Page After
C. Section Expert – Color Tab - New Page Before
D. Refresh report and this should clear the blank section

Q46. Which of the following will allow you to place a watermark behind data displayed in your report?

A. This formatting method does not exist in Crystal Reports
B. Export the report to Excel and re import into Crystal
C. Underlay
D. Print the report and scan the watermark in through Photoshop
E. Use Acrobat Reader

Q47. You want the group and its contents to start on a new page. From the Section Expert select the Group Paging Tab and tick?

A. Free-From placement
B. Reset Page Number
C. Keep Together
D. New Page Before

Q48. Which of the following will allow you to format a group with multiple columns?

A. Format multiple columns can only be applied to the details section
B. Right click the group, select Section Expert from the pop-up menu and tick the checkbox, which read Format multiple columns
C. From the toolbar select report and Group Expert, select format multiple columns under the Group Header
D. From the toolbar select report and Section Expert, select format multiple columns under the Group Header

Q49. How can you access the Section Expert? *(Multiple Answers)*

A. From the Menu Bar select Report –Section Expert

B. Select the Section Expert icon from the toolbar
C. Right-click the grey area of the section and select Section Expert from the drop-down menu
D. Choose format from the Menu Bar and select Section Expert.

Q50. What are the advantages of using the Highlighting Expert? *(Multiple Answers)*

A. Knowledge of Crystal or Basic Syntax is not required
B. The same format can be used for several fields
C. Formatting based on font style, color, border, background, and value can be applied to several fields
D. Fields can be formatted based on the results of other fields

Q51. Which of the following will allow you to apply formatting applicable to field 1 to field 2?

A. Format Painter

B. Highlighting Expert
C. Format Ruler
D. Format Page
E. Report Options

Q52. **Using the Highlighting Expert, you want the Customer_ No to be assigned the color blue if the client city equals 'London'. Which of the following apply?**

A. Right-click Customer_No and select Highlighting Expert from the drop-down menu, select New, in the Item Editor section select Value of city is equal to 'London' then Font color = blue and click ok

A. Right-click Customer_No and select Highlighting Expert from the drop-down menu, select New, in the Item Editor section select Value of Customer_No is equal to 'London' then Font color = blue and click ok

B. Right-click Customer_No and select Highlighting Formatter from the drop-down menu, select New, in the Item Editor section select Value of city is equal to 'London' then Font color = blue and click ok

C. Right-click Customer_No and select Highlighting Section from the drop-down menu, select New, in the Item Editor section select Value of city is equal to 'London' then Font color = blue and click ok

Q53. **You have applied a format to the Customer_No field using the Highlighting Expert. You want to add additional formatting to the same field, where Customer_No will equal red if Country is USA. Which of the following apply?**

A. Right-click Customer_No and select Highlighting Expert from the drop-down menu, click New, in the Item Editor section select Value of Customer_No is equal to 'USA' then Font color = red and click ok

B. Right-click Customer_No and select Highlighting Formatter from the drop-down menu, click New, in the Item Editor section select Value of country is equal to 'USA' then Font color = red and click ok

C. Right-click Customer_No and select Highlighting Expert from the drop-down menu, click New, in the Item Editor section select Value of country is equal to 'USA' then Font color = red and click ok

D. Additional Formatting cannot be applied to a field which has already been formatted using the Highlighting Expert

Q54. What does the Format Painter allow you to do?

A. Format Painter allows you to delete objects.
B. Format Painter allows you to copy formatting from one field to another, by Right-clicking the field you want use and selecting Format Painter and clicking the field you want to apply the formatting to
C. Format Painter allows you to copy formatting from one field to another
D. Format Painter allows you to copy numbers only, by Right-clicking the field you want use and selecting Format Painter and clicking the field you want to apply the formatting to
E. Highlight changes

Q55. Objects placed in the Report Header will print once at the beginning of the report.

F. True
A. False

Q56. You want to format a report to insert a blank line after every 5th record in the detail section. Which of the following formulas should you apply?

A. Insert a detail section b, Remainder (RecordNumber) <> 0 should be added to the detail section b of the report, under the Section Expert and under the Common Tab and suppress x+2 section
B. Insert an additional detail section under the details section (detail section b), place the following formula Remainder (RecordNumber, 5) <> 0 in detail section b of the report, within the Section Expert – Common Tab highlight details sections(b) and click the suprpress formula button and select formula created
C. Insert a detail section b, Remainder (RecordNo = 5) <> 0 should be added to the detail section b of the report, under the Section Expert and under the Common Tab and suppress x+2 section
D. Insert a detail section b, Remainder (RecordNumber) = 5 should be added to the detail section b of the report, under the Section Expert and under the Common Tab and suppress x+2 section

Q57. **You want to create a report template without a data source. Which of the following methods apply?**

A. Select File - New - Blank Report, select a data source from the Database Expert, click the ok button and insert a Template Field Object, format the template field object accordingly

B. Select Report – Report Template, do not select a data source from the Database Expert, click the cancel button and insert a Template Field Object, format the template field object accordingly

C. Select Report – Report Template, select a data source from the Database Expert, click the ok button and insert a Template Field Object, format the template field object accordingly

D. Select File - New - Blank Report, do not select a data source from the Database Expert, click the cancel button and insert a Template Field Object, format the template field object accordingly

Q58. **You want to add a title to a report. Which of the following apply?** *(Multiple Answers)*

A. Insert - Text Object and enter title
B. Report - Text Object and enter title
C. Text - Text Object and enter title
D. View - Text Object and enter title
E. Select Report Title from the Field Explorer – Special Fields

Q59. **You want to increase the height and width of a Text Object. Which of the following apply?**

A. Right-click the Text Object, from the dropdown menu select Grid, enter the Height and Width required

B. Right-click the Text Object, from the dropdown menu select Size and Position, enter the Height and Width required

C. Right-click the Text Object, from the dropdown menu select X and Y, enter the Height and Width required

D. Right-click the Text Object, from the dropdown menu select Height and Width, enter required size

Q60. **You want to insert a Microsoft PowerPoint Presentation into your current report. What procedure should you follow?**

A. This function does not exist
B. Select File – insert – Microsoft PowerPoint Presentation file – locate file and click ok
C. Select Insert – File - – locate file and click ok
D. Select Insert – OLE Object – Highlight Microsoft PowerPoint Presentation - select Create from File, browse to locate the file

Q61. **Which of the following can be inserted into a report? (Multiple Answers)**

A. Cross-Tab
B. Text Object
C. Flash
D. OLAP Grid, Pictures, Map
E. Subreport
F. OLE Object, Box
G. Line, Chart

Q62. **How would you add a logo to your report?**

A. Select Insert - Picture, from the dialog box locate the logo and click OK, place the logo in a required section of your report
B. Select Insert - Object, from the dialog box locate the logo and click OK, place the logo in a required section of your report
C. Select View - Insert - Picture from the toolbar, locate the logo and click OK, place the logo in your report
D. Select Report - Insert - Picture from the toolbar, locate the logo and click OK, place the logo in your report

Q63. **Which of the following appear under OLE Object insertion? (Multiple Answers)**

A. Adobe Acrobat Document
B. Adobe Photoshop image
C. Bitmap image
D. Media Clip
E. Microsoft Excel Chart

F. Microsoft PowerPoint Presentation

G. Business Objects Documentation

Q64. **You have grouped your report by Customer_No and you want to add the text 'Customer No ' beside the database field within the group. Which of the following apply?**

A. Select Insert Text Object, places the Text Object in the group, enter Customer No , drag the group into the Text Object after the Customer No :

B. Select Insert Text Box, insert Text Box in the group, enter Customer No , drag the group into the Text Box after the Customer No :

C. Select Insert Object, insert object in the group, enter Customer No , drag the group into the object after the Customer No :

D. Select Insert Textbook Object, insert Textbook object in the group, enter Customer No , drag the group into the Textbook object after the Customer No :

Q65. **You want to align all fields in the details section. Which method applies?**

A. Right-click the grey area of the details (b) section and select Align Tops

B. Right-click the grey area of the details (b) section and select Align all objects

C. Right-click the grey area of the details (b) section and select 'Select All Section Objects, Right-click highlighted fields and select Align- Tops or as required

D. Right-click the grey area of the details (b) section and select Tops

Q66. **You want to set all fields in the Details Section to the same size. Which method applies?**

A. Right-click the grey area of the details (b) section and select Same Size

B. Right-click the grey area of the details (b) section and select section Same Size

C. Right-click the grey area of the details (b) section and select Align Same Size

D. Right-click the grey area of the details (b) section and select 'Select

All Section Objects, Right-click highlighted fields and select Size
– Same Size

Q67. **You want to apply a color to a field's border within the Format Editor, which of the following apply?**

A. Border Tab
B. Common Tab
C. Paragraph Tab
D. Hyperlink Tab
E. Font Tab

Q68. **You want to insert a section below your existing Group Header #2. Which of the following methods apply?** *(Multiple Answers)*

A. Select Format – Section Expert – highlight Group Header #2 and select Insert
B. Select File – Section Expert – highlight Group Header #2 and select Insert
C. Select Report – Section Expert – highlight Group Header #2 and select Insert
D. Right-click the grey section of GroupHeader#2 and select Insert Section below

Q69. **You have two sections within your report GroupHeader#3a and GroupHeader#3b. You want to combine both sections together. Which of the following apply?** *(Multiple Answers)*

A. Select Report – Section Expert – highlight Group Header #3a and select merge
B. Right-click the grey section of GroupHeader#3a and select merge section below
C. Select File – Section Expert – highlight Group Header #3a and select merge
D. Right-click the grey section of GroupHeader#3b and select merge section below

Q70. **How would you resize the details section of your report?**

A. You can double space the report by right-clicking the Details Section and select double spacing from the pop-up menu
B. You can double space the report by right-clicking the Details Section and select space and a half
C. Place the cursor on the bottom line of the detail section, the cursor will display two lines with an up and down arrow, pull down or up according to your requirements

Q71. Which one of the following can be used to set dynamic locations for graphics within your report?

A. Dynamic Graphic Location
B. Active Graphic Location
C. ActiveX Graphic Location
D. XML Graphic Location

Q72. Dynamic Graphic Location Conditional formatting can be applied to a Blob Field?

A. True
B. False

Q73. Where is Allow Field Clipping activated?

A. Within the Format Editor – Border Tab – Click Customize – Number Tab and Check the Allow Field Clipping checkbox
B. Within the Format Editor – Number Tab – Click Customize – Number Tab and Check the Allow Field Clipping checkbox
C. Within the Format Editor – Currency Tab – Click Customize – Number Tab and Check the Allow Field Clipping checkbox
D. Within the Format Editor – Common Tab – Click Customize – Number Tab and Check the Allow Field Clipping checkbox

Q74. Which of the following are the three tabs available via the Section Expert? *(Multiple Answers)*

A. Common
B. Font
C. Color
D. Paging

Q75. **The Paging Tab provides the functionality to set which of the following?** *(Multiple Answers)*

A. New page after
B. New page before
C. Reset page number after
D. Orientation
E. None of the above

Q76. **Which of the following options exist within the Section Expert under the color tab?**

A. Font Color
B. Background color
C. Object Color

Q77. **A barcode can be applied to a date time field**

A. True
B. False

Q78. **You want to apply a barcode to your field. Which of the following apply?**

A. Right-click the field and select change to barcode from the dropdown menu and select barcode type
B. Select the field, from the menu bar select Format - change to barcode – select barcode type
C. Select the View, from the menu bar select Format - change to barcode – select barcode type
D. Select the Format, from the menu bar select Format - change to barcode – select barcode type

Q79. **Which of the following are the barcode types available?** *(Multiple Answers)*

A. Code49
B. Code39
C. Code59
D. Code39 Full ASCII
E. Code59 Full ASCII

Q80. You have placed the Report Title from the Field Explorer onto your report; the report title is empty. Which of the following explains the absence of the title?

A. The Report Info title is empty
B. The Report title is empty
C. The Summary Info title is empty
D. The Info title is empty

Q81. You have created a group based on sales representative, you have also created a summary of sales for each representative; you want to highlight the background of the details section in red for all sales representatives who have not met the sales target of £2000. Which of the following apply?

A. Within the Details section of the Section Expert, select the color tab and click the formula button, and enter if sum(Sales.SalesAmount},{Sales.SalesRep}) < 2000 then crred else crnocolor.
B. Within the Details section of the Section Expert, select the font tab and click the formula button, and enter if sum(Sales.SalesAmount},{Sales.SalesRep}) < 2000 then crred else crnocolor.
C. Within the Details section of the Section Expert, select the Group tab and click the formula button, and enter if sum(Sales.SalesAmount},{Sales.SalesRep}) < 2000 then crred else crnocolor.
D. Within the Details section of the Section Expert, select the paging tab and click the formula button, and enter if sum(Sales.SalesAmount},{Sales.SalesRep}) < 2000 then crred else crnocolor.

Q82. You want to place the Customer Name field from the Field Explorer within your report. Which of the following methods apply?

A. From the menu select View – Field Explorer – Expand the Database Fields section and also expand the table required – with the cursor select the Customer Field holding the left section of the mouse down and move the field to the section within the report
B. From the menu select Report – Field Explorer – Expand the Database Fields section and also expand the table required – with the cursor select the Customer Field holding the left section of the mouse down and move the field to the section within the report

C. From the menu select Format – Field Explorer – Expand the Database Fields section and also expand the table required – with the cursor select the Customer Field holding the left section of the mouse down and move the field to the section within the report
D. From the menu select Database – Field Explorer – Expand the Database Fields section and also expand the table required – with the cursor select the Customer Field holding the left section of the mouse down and move the field to the section within the report

Adding Graphical Elements

Q1. **You want to draw a line under your Title. Which of the following apply?**

    A. select Report - Insert - Line, use the pencil icon to click the start point of the line and drag to the endpoint

    B. select Insert - Line, use the pencil icon to click the start point of the line and drag to the endpoint

    C. select Format - Insert - Line, use the pencil icon to click the start point of the line and drag to the endpoint

    D. select View - Insert - Line, use the pencil icon to click the start point of the line and drag to the endpoint

Q2. **You want to insert a box within your report. Which of the following apply?**

    A. select Report - Box, use the pencil icon to click the start point of the box and drag to the size required

    B. select Format - Box, use the pencil icon to click the start point of the box and drag to the size required

    C. select View - Box, use the pencil icon to click the start point of the box and drag to the size required

    D. select Insert - Box, use the pencil icon to click the start point of the box and drag to the size

Q3. **Which of the following formats can be applied to a Line?**
(Multiple Answers)

    A. Style
    B. Width
    C. Colour
    D. Size
    E. Drop Shadow

Q4. **Which of the following formats can be applied to a Box?**
(Multiple Answers)

    A. Style
    B. Width

C. Colour
D. Size
E. Drop Shadow
F. Rounding

Q5. **Text can be typed directly into a box**

A. True
B. False

Q6. **What style options are available to a Line?** *(Multiple Answers)*

A. None
B. Single
C. Dashed
D. Dotted

Q7. **Which of the following options are available when setting the width of a line?** *(Multiple Answers)*

A. Hairline
B. Hairline 2.5pt
C. Hairline 1pt
D. Hairline 3.5pt

Q8. **You want to move a line. Which of the following apply?**

A. Click the line and drag to required position
B. Double-click the line, this will move the line to the required position

Q9. **You want to add a picture to your report. Which of the following apply?**

A. Select Insert - .bmp - Browse and select picture required and click open
B. Select Insert - File - Browse and select picture required and click open
C. Select Insert - Object - Browse and select picture required and click open

D. Select Insert - Picture - Browse and select picture required and click open

Q10. You want to add a dynamic image. Which of the following apply?

A. Select Insert - Picture – Browse picture and click open - place picture in report as required, right-click the picture and select format graphic, click the Picture tab - click the formula button beside the Graphic Location and enter the file location

B. Select Format -Insert - Picture – Browse picture and click open - place picture in report as required, right-click the picture and select format graphic, click the Picture tab - click the formula button beside the Graphic Location and enter the file location

C. Select Report - Insert - Picture – Browse picture and click open - place picture in report as required, right-click the picture and select format graphic, click the Picture tab - click the formula button beside the Graphic Location and enter the file location

D. Select File - Insert - Picture – Browse picture and click open - place picture in report as required, right-click the picture and select format graphic, click the Picture tab - click the formula button beside the Graphic Location and enter the file location

Q11. Which of the following image files are supported by Crystal Reports 2008?

A. TIFF
BITMAP
JPEG
PNG
META
GIF

Apply Section Formatting

Q1. **You want to apply different page orientations to different sections of your report. Which of the following methods apply?**

A. Right click the report section and select the Format Section Expert, click the page tab and under orientation select landscape or portrait, follow the same procedure to apply the required orientation to the next section

B. Right click the report section and select the Section Expert, click the paging tab and under orientation select landscape or portrait, follow the same procedure to apply the required orientation to the next section

C. Right click the report section and select the section format expert, click the paging tab and under orientation select landscape or portrait, follow the same procedure to apply the required orientation to the next section

D. Right click the report section and select the Section Expert, click the common tab and under orientation select landscape or portrait, follow same procedure to apply the required orientation to the next section

Q2. **You want to widen a section within your report. Which of the following methods apply?**

A. Move the cursor to the line at the bottom section, click and drag to required position when the cursor turns to 2 lines with an upward and downward arrow.

B. Move the cursor to the line at the bottom section, and double -click a dialog box will appear, enter the size required.

Q3. **You want to view the full name of a section. Which of the following apply?**

A. Right-click the section in design view and select show long section names

B. Right-click the section in preview view and select show long section names

C. Right-click the section in design view and select show section names

D. Right-click the section in preview view and select shown section names

Q4. **You have right-clicked the Details section and selected Section Expert. Which of the following formatting options will be available?** *(Multiple Answers)*

A. Suppress (No Drill Down)
B. Insert Line
C. Delete Last Line
D. Arrange Lines
E. Fit Section
F. Format Section

Q5. **Which of the following will reduce the size of the section to the height of the lowest object in the section?**

A. Suppress (No Drill Down)
B. Insert Line
C. Delete Last Line
D. Arrange Lines
E. Fit Section
F. Format Section

Q6. **Which of the following ensures all horizontal lines are arranged in the section?**

A. Suppress (No Drill Down)
B. Insert Line
C. Delete Last Line
D. Arrange Lines
E. Fit Section
F. Format Section

Q7. **Which of the following formatting options is applicable to all sections of a report?**

A. Section Expert
B. Format Expert
C. Report Expert
D. Option Expert

Q8. **Which of the following options will still allow the user to drill down into the next sections when set?**

A. Drill Down Only
B. Suppress (Drill Down OK)
C. Hide (Drill Down OK)
D. Hide (No Drill Down)
E. Keep Together

Q9. **Which of the following options when ticked prevents page breaks from being inserted within the section?**

A. Drill Down Only
B. Suppress (Drill Down OK)
C. Hide (Drill Down OK)
D. Hide (No Drill Down)
E. Keep Together

Q10. **You want to prevent objects within the details sections from being moved. Which of the following options apply?**

A. View Only
B. Relative Position Only
C. Suppress Movement
D. Lock Position and Size

Q11. **Which of the following options are available within the Common Tab of the Section Expert?** *(Multiple Answers)*

A. View Only
B. Relative Positions
C. Suppress Movement
D. Read Only
E. Suppress (Drill Down OK)
F. Print at bottom of page
G. Keep Together
H. Underlay following section
I. Format multiple columns

Q12. By default how many tabs are available under the Section Expert? *(Multiple Answers)*

A. Common Tab
B. Paging Tab
C. Report Tab
D. Background Tab
E. Color Tab

Q13. The paging tab can be used to apply which of the following to a section? *(Multiple Answers)*

A. New Page Before
B. New Page After
C. Suppress Page
D. Hide Page

Q14. Where can you apply background color to a report section?

A. Common tab
B. Font tab
C. Background tab
D. Color tab
E. Layout Tab

Q15. Which of the following sections can be set when multiple columns has been selected? *(Multiple Answers)*

A. Detail Size
B. Font tab
C. Gap between details
D. Printing Direction

Q16. You want a new page after every 10 records in the details section. Which of the following apply?

A. From the common tab within the Section Expert of the details section, tick visible records and enter 10
B. From the font tab within the Section Expert of the details section, tick visible records and enter 10
C. From the reports tab within the Section Expert of the details section, tick visible records and enter 10

D. From the paging tab within the Section Expert of the details section, tick visible records and enter 10

Q17. **You have applied a New Page Before to the first groupheader within your report; this has created a first blank page in preview. What can you do to remove this blank page?**

A. Apply Not OnFirstRecord within the formula section of the New Page Before within the Section Expert
B. Apply Not OnLastRecord within the formula section of the New Page Before within the Section Expert
C. Apply Not OnFirstPage within the formula section of the New Page Before within the Section Expert
D. Apply Not OnLastPage within the formula section of the New Page Before within the Section Expert

Q18. **You want each group within your report to start with a new page number, which of the following apply?**

A. Within the Section Expert click the common tab and select New Page Before, apply the conditional formatting and select reset page number after
B. Within the Section Expert click the paging tab and select New Page Before, apply the conditional formatting and select reset page number after
C. Within the Section Expert click the report tab and select New Page Before, apply the conditional formatting and select reset page number after
D. Within the Section Expert click the background tab and select New Page Before, apply the conditional formatting and select reset page number after

Q19. **Which of the following are applicable to the Page Footer Only?** *(Multiple Answers)*

A. Clamp Page Footer
B. Reserve Minimum Page Footer
C. Read Only
D. Relative Position

Q20. Which of the following settings will size the page to content when previewed?

A. Clamp Page Footer
B. Reserve Minimum Page Footer
C. Read Only
D. Relative Position

Q21. You have selected format with Multiple Columns. Which one of the following additional tabs will appear?

A. Common
B. Layout
C. Paging
D. Color

Q22. You have inserted a Cross-Tab within your report however when you preview the page numbers do not span across the horizontal pages. How can you rectify this?

A. Refresh the report
B. Right-click the cross-tab and Pivot Cross-Tab
C. Insert a Horizontal Page Number from the Special Fields section of the Field Explorer
D. Horizontal Page Numbers do not exists within Crystal Reports

Q23. Which of the following options are available under the Field Explorer within the Special Fields? *(Multiple Answers)*

A. File Path and Name
B. Record Number
C. Modification Date
D. Modification Time
E. Report Comments
F. Report Title

Q24. Which of the following options are available under the Field Explorer within the Special Fields? *(Multiple Answers)*

A. Page N of M
B. Group Selection formula

C. Print Time Zone

D. Print Date

E. Content Locale

Q25. **You want to display the name of the report Author on your report, which special field applies?**

A. Author

B. File Author

C. Report Author

D. User Name

Q26. **You want to display the number of groups within your report, which special field applies?**

A. Group Count

B. Group Path Number

C. Group Selection Number

D. Group Number

Q27. **You want to display the last time your report was changed, which special field applies?**

A. Modification Date

B. Modification Time

C. Print Date

D. Print Time

E. Modification Date Time

Q28. **You have applied a new page after to the group header within your report the first page is blank which functions will rectify this?**

A. On LastPage

B. Not OnLastPage

C. Not OnFirstPage

D. Not OnFirstRecord

E. <> Null

F. Not OnLastRecord

Create a Chart

Q1. **Which of the following data categories can be used to create a Chart?** *(Multiple Answers)*

A. Group Summaries and Subtotals
B. Data in the details section
C. Data within an existing Cross-Tab
D. Formulas with Evaluation Time Functions (WhilePrintingRecords)
E. Formulas
F. Running Totals
G. OLAP data

Q2. **Which of the following statements are untrue about Charts?** *(Multiple Answers)*

A. Charts cannot be created from Running Totals
B. Charts cannot be created from Cross-Tabs
C. Charts cannot be deleted once created
D. Charts can only be created when summaries exists within the report

Q3. **You place your Chart in the Report Header. What will the Chart represent?**

A. All report data
B. Group Data only
C. Group Header data only
D. Group Footer data only

Q4. **Given the data below, which Bar Chart Type will be most appropriate?**

Student Count	County	Month	Year
8000	London	May	2008
789	Birmingham	May	2008
1240	Essex	May	2008
516	Dublin	May	2008
690	Coventry	May	2008
364	London	June	2008
4299	Birmingham	June	2008
554	Essex	June	2008
5553	Dublin	June	2008
5523	Coventry	June	2008
1510	Kent	June	2008
1067	Bolton	June	2008

A. Percent bar Chart
B. Stacked
C. Stacked bar Chart
D. Side by side bar Chart

Q5. **How can a chart be saved as a template?**

A. Right-click the Chart and select Chart Expert, within the Data Tab select save as template, assign a name and save the Chart within the user defined folder
B. Right-click the Chart and select save as template, assign a name and save the Chart within the user defined folder
C. Right-click the Chart and select Chart Expert, within the template tab select save as template, assign a name and save the Chart within the user defined folder
D. Right-click the Chart and select save as template, assign a name and save the Chart within the user defined folder
E. None of the above, Charts cannot be saved as templates

Q6. **Which of the following are Chart Types within the Chart Expert?** *(Multiple Answers)*

A. Bar, Line, Area
B. Numeric Axis, Gauge, Gantt
C. Pie, Doughnut, 3D Riser
D. Pie, Doughnut, 2D Riser
E. XY Scatter, Stock, Radar, Bubble
F. Y Scatter, OLAP
G. Funnel, Histogram

Q7. **Chart data is incompatible. Which of the following will occur?**

A. The OK button will be greyed out
B. An Exit Chart Expert dialog box will appear
C. Return to Chart Expert will be displayed
D. The Chart Expert will automatically choose the most appropriate Chart Type

Q8. **Launching the Chart Expert involves which of the following?**

A. Choose Report, insert Chart, select the Chart Type, select the Data Tab for data type and Chart (group, advance, OLAP, or Cross-Tab) and text tab for titles
B. Choose insert Chart, a Chart placeholder will automatically appear, place this Chart in the required section, the Chart Expert will appear, select the Chart Type, data, axes and other customized settings
C. Choose Section Expert, insert Chart, select the Chart Type, select the Data Tab for data type and Chart (group, advance, OLAP, or Cross-Tab) and text tab for titles
D. Choose Report Expert, insert Chart, select the Chart Type, select the Data Tab for data type and Chart (group, advance, OLAP, or Cross-Tab) and text tab for titles

Q9. **Within the Chart Expert, which of the following tabs are available?** *(Multiple Answers)*

A. Type, Color Highlight, Data, Options

B. Axes, Text, Legend
C. Options, Highlight Expert
D. General, Templates, Data Grid, Selected Item
E. Axes, Text

Q10. **Given the chart below you want to change the colour of the London bar to 'Red'. Which of the following apply?**

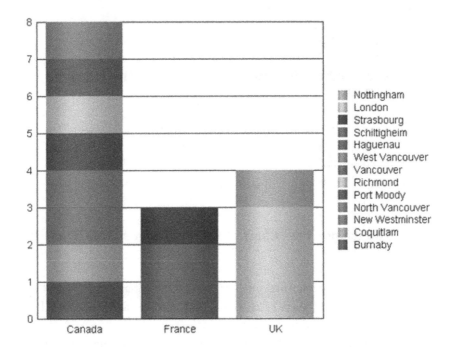

A. Right-click Chart and select Chart Expert, select the Highlighting Tab and select New where value of City is equal to 'London' – format color is Red
B. Right-click Chart and select Chart Expert, select the Format Tab and select New where value of City is equal to 'London' – format color is Red
C. Right-click Chart and select Chart Expert, select the Color Expert Tab and select New where value of City is equal to 'London' – format color is Red
D. Right-click Chart and select Chart Expert, select the Color Highlight Tab and select New where value of City is equal to 'London' – format color is Red

Q11. Creating an Advanced Chart does not require a group within the report or field details.

    A. True
    B. False

Q12. You have created a Chart as displayed below. You want to change the order of the cities as follows: Vancouver, London, Strasbourg, Which of method is applicable?

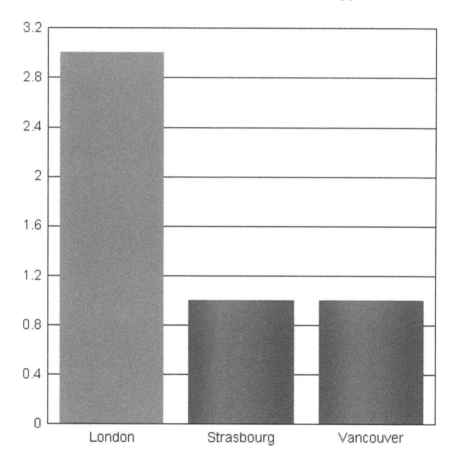

    A. Right-click Chart and select Chart Expert, select the Specified Data Tab and highlight the county field in the On Change of section, click the order button and select Specified Order from the Common Tab and, click the Specified Order tab and select Vancouver, London, Strasbourg from the Named Group and click ok

B. Right-click Chart and select Chart Expert, select the Group Data Tab and highlight the county field in the On Change of section, click the order button and select Specified Order from the Common Tab and, click the Specified Order tab and select Vancouver, London, Strasbourg from the Named Group and click ok

C. Right-click Chart and select Chart Expert, select the Data Tab and highlight the county field in the On Change of section, click the order button, select Specified Order from the Common Tab, click the Specified Order tab and select Vancouver, London, Strasbourg from the Named Group and click ok

D. Right-click Chart and select Chart Expert, select the Options Tab and highlight the county field in the On Change of section, click the order button and select Specified Order from the Common Tab and, click the Specified Order tab and select Vancouver, London, Strasbourg from the Named Group and click ok

Q13. Which of the following are Chart Expert Data Layout Types? *(Multiple Answers)*

A. Advanced
B. Group
C. Cross-Tab
D. OLAP

Q14. How would you load a stored template into your existing Chart?

E. Right-click the Chart and select Chart Expert, and Load Template, select the required template from the categories section and click ok

F. Right-click the Chart and select Load Template, select the required template from the categories section and click ok

G. Right-click the Chart and select Chart options and Load Template, select the required template from the categories section and click ok

H. Right-click the Chart and select Template, select the required template from the categories section and click ok

Q15. **How would you display the top 10 product sales within your advanced Chart?**

A. Select Report from the Menu Bar and select Group Sort Expert and select Top N based on sum product sales where N equals 5 and click ok

B. Right-click Chart and select Chart Expert, select the Text tab and highlight the county field in the On Change of section, click the Top N button and select Top N based on sum product sales where N equals 5 and click ok

C. Select Report from the Menu Bar and select record sort expert and select Top N based on sum product sales where N equals 5 and click ok

D. Right-click Chart and select Chart Expert, select the Data Tab and highlight the Product field in the On Change of section of the Chart Expert, click the Top N button and select Top N based on sum of product sales where N equals 5 and click ok

Q16. **You want to change the color of a bar within your Chart. Which of the following apply?**

A. Right-click the Chart and select Chart Expert and format series riser, choose the required color from the Foreground color

B. Right-click the bar and select format series riser and choose the required color from the fore ground color

C. Right-click the Chart and select format Chart Expert and choose the required color from the fore ground color

D. Right-click the Chart and select format bar and choose the required color from the fore ground color

Q17. **You have double-clicked a Chart to drill-down, but nothing happens. Which one of the following reasons applies?**

A. You have placed the Chart in the details section

B. The Chart contains a formula

C. The Chart is based on a running total

D. The Chart is an advanced Chart and therefore cannot be drilled-down into

Q18. **Data for a chart can be selected from which one of the following sections?**

A. Type
B. Fields
C. Axes
D. Data Options
E. Data

Q19. **Within your report you have groups and summaries. Which Chart layout type will be most appropriate?**

A. Advanced Chart
B. Group Chart
C. OLAP Chart
D. Cross Tab Chart

Q20. **You want to change your Chart bar to a vertical display. What should you do?**

A. Right-click the Chart and select Chart options, from the appearance tab select Vertical
B. Right-click the Chart and select Chart Expert, from the Type tab select Vertical
C. Right-click the Chart and select Rotate Chart, from the drop-down menu select Vertical
D. Right-click the Chart and format Chart Expert, from the Common Tab select Vertical

Q21. **Drill-down charts are only applicable when a group and summaries exist within the report.**

A. True
B. False

Q22. **You want to apply depth to your pie chart. Which of the following apply?**

A. Right-click the Chart and select Chart Expert, from the Axes tab select Use depth effect
B. Right-click the Chart and select Chart Expert, from the Depth tab select Use depth effect

C. Right-click the Chart and select Chart Expert, from the Type tab select Use depth effect
D. Right-click the Chart and select Chart Expert, from the Text tab select Use depth effect

Q23. You can see the result of your Chart in preview mode.
E.
A. True
B. False

Q24. Your Chart appears in black and white. Which of the following settings has been applied?

A. Chart Expert – Color –Chart Color: Black and White
B. Chart Expert – Color Highlights –Chart Color: Black and White
C. Chart Expert – Data –Chart Color: Black and White
D. Chart Expert – Options –Chart Color: Black and White

Q25. Which of the following tabs within the Chart Expert will be unavailable when you select any of the following Chart Types: Pie, Doughnut, Gantt or Funnel?

A. Axes
B. Options
C. Text
D. Color Highlight

Q26. Which of the following settings can be implemented under the Axes tab? *(Multiple Answers)*

A. Major. Minor
B. Group or Data axis
C. Auto Scale
D. Auto Range

Q27. You want to assign a specific colour (Blue) to your bar Chart, if the city equals 'London'. Which method applies?

A. Within the Chart Expert, select the Options Tab and click the new button, the city field will appear, set the value to London and select the color Blue and click ok

B. Within the Chart Expert, select the Color Highlights tab and click the new button, the city field will appear, set the value to London and select the color Blue and click ok
C. Within the Chart Expert, select the Axes tab and click the new button, the city field will appear, set the value to London and select the color Blue and click ok
D. Within the Chart Expert, select the Data Tab and click the new button, the city field will appear, set the value to London and select the color Blue and click ok

Q28.   How would you add a title to your chart?

A. Within the Chart Expert, select the Text tab and enter your new title
B. Within the Chart Expert, select the Data Tab and enter your new title
C. Within the Chart Expert, select the Options Tab and enter your new
D. Within the Chart Expert, select the File tab and enter your new title

Q29.   Which Chart is most suitable for time display over a period?

A. Pie Chart
B. 3 –D Surface
C. Bar Chart
D. Doughnut
E. Line

Q30.   You want to add data labels to your chart which of the following apply?

A. Chart Expert – Options – Data Points, Show Label
B. Chart Expert – Labels – Data Points, Show Label and Show Value
C. Chart Expert – Values – Data Points, Show Label and Show Value
D. Chart Expert – Text – Data Points, Show Label and Show Value

Q31. You do not want the Chart Legend to appear. Which of the following methods will remove it?

A. Within Chart Expert – Options Tab – Legend, uncheck Show Legend
B. Within Chart Expert – Legend Tab – Legend, uncheck Show Legend
C. Within Chart Expert – Text Tab – Legend, uncheck Show Legend
D. Within Chart Expert – Placement Tab – Legend, uncheck Show Legend

Q32. You have right-clicked your Chart in Preview Mode. Which of the following options will be available? *(Multiple Answers)*

A. Auto-Arrange Chart
B. Zoom in
C. Load Template
D. Save as Template
E. Select Mode
F. Format Background
G. Format Foreground
H. Size and Position

Q33. Which of the following three options allow the user to choose when the Advanced Chart starts a new block? *(Multiple Answers)*

A. On change of
B. For each record
C. For all records
D. For group

Q34. Which of the following settings are available within the Option Tab of the Chart Expert? *(Multiple Answers)*

A. Layout
B. Chart Color
C. Legend
D. Customize Settings
E. XY displays

Q35. Which Chart represents the percentage for each item?

A. Line
B. 3-D riser
C. Pie Chart
D. 3 –D Surface

Q36. What does the figure in the middle a doughnut chart represent?

A. The Subtotal for the groups
B. The Grand Total for the whole report
C. The summary for the groups
D. The summary for the page headers

Q37. Which of the following is not a Chart Type in Crystal? *(Multiple Answers)*

A. Rule Chart
B. Line Chart
C. Pie Chart
D. 2-D Riser
E. Gauge
F. Bubble

Q38. Group Charts represent data that already exists in one of the following:

G. Subtotals and summary fields in the detail section
H. Subtotals and summary fields in the Report Header
I. Subtotals and summary fields in the page header
J. Subtotals and summary fields in the Group Header or Group Footer

Q39. The Data Tab within the Chart Expert provides which of the following functionalities? *(Multiple Answers)*

A. Denote A Group Chart Layout
B. Denote An Advanced Chart Layout
C. Denote A Cross-Tab Chart Layout
D. Denote An OLAP Layout
E. Denote Chart Type, Pie, Bubble Etc

Q40. **You right-click a Stacked Bar Chart and select Chart Options. Which of the following settings will appear?** *(Multiple Answers)*

A. Appearance
B. Titles
C. Data Labels
D. Gridlines
E. Axes
F. Legend

Q41. **You have right-clicked a bar within your Bar Chart. Which of the following options will be available?** *(Multiple Answers)*

A. Format series riser
B. Chart options
C. Series option
D. Trend lines
E. Lines

Q42. You want to change your Bar Chart to a Funnel Chart Type; the data is not compatible with this Chart Type. What will happen?

A. The ok button will be greyed out
B. The Chart Data and Type dialog box will appear indicating the data does not fit the Chart Type selected
C. The Chart will be presented as a blank section
D. All Chart Types are compatible will all data
E. The Chart Selection Expert will appear providing an alternative option

Q43. **Which options will appear when an incorrect Chart Type is selected for data in your report?** *(Multiple Answers)*

A. Continue
B. Return to Chart Expert
C. Automatically choose the most appropriate Chart Type
D. None of the above

Q44. **Each Chart Type within the Chart Expert provides a Chart functional description.**

A. True
B. False

Q45. **Copy Smart Tag allows the user to do which of the following?**

A. Allows the user to copy a chart into any Office XP application
B. Allows the user to delete a locked chart
C. Allows the user to delete a read only chart

Q46. **Which of the following are Trend lines options?** *(Multiple Answers)*

A. Mean
B. Polynomial
C. Power Regression
D. Quadratic Regression
E. Log Regression
F. Corrected Lines

Q47. **Which of the following Office XP applications work with Copy Smart Tag?** *(Multiple Answers)*

A. MS Word
B. MS Excel
C. Outlook

Q48. **You right-click a stacked chart. Which of the following options will be greyed out?**

A. Appearance
B. Titles
C. Data Labels
D. Gridlines
E. Multi Axes
F. Legend

Q49. **Which of the following options are available under Format Series Riser?**

A. Fill
B. Border
C. Data
D. Axes

Apply Report Templates

Q1. **How would you apply formatting to a template without connecting to a data source?**

A. Create Template Detail
B. Create a Template Field Object
C. Create Report | Template Details
D. Template Layout Function with data source set to Log off model

Q2. **Where do the sample templates reside by default?**

A. \Program Files\Business Objects\Crystal Reports 12.0\Templates\ en
B. \Program Files\Crystal Reports 11\Templates
C. \Program Files\Business Objects\Sample\Crystal Reports 11\ Templates
D. \Program Files\Business Objects\Sample Reports\Crystal Reports 11\Templates

Q3. **How would you apply a template to a report?** *(Multiple Answers)*

A. Choose Report | Template Expert | from the Template Expert box select the template required and apply to the current report
B. Click the Template Expert icon and select the sales template and click ok
C. Choose Format | Template Expert | from the Template Expert box select the template required and apply to the current report
D. Choose View | Template Expert | from the Template Expert box select the template required and apply to the current report

Q4. **You notice a report you saved as a template does not appear under available templates, which of the following explains this?**

A. Verify the database connection
B. Press F5 within the available templates section
C. Open the report from the templates folder – select file, Summary Info and enter a title
D. Press F5 within windows explorer

Q5.   You want to create a report using an existing template, which method applies?

A. Select Report - from the Standard Report Creation Wizard choose the tables - create links – choose fields to display – group by – apply summaries and filters if required and select a template from the available templates and click finish

B. Select Template - from the Standard Report Creation Wizard choose the tables - create links – choose fields to display – group by – apply summaries and filters if required and select a template from the available templates and click finish

C. Select Format - from the Standard Report Creation Wizard choose the tables - create links – choose fields to display – group by – apply summaries and filters if required and select a template from the available templates and click finish

D. Select New – Using the Report Creation Wizard select Standard Report choose the tables - create links – choose fields to display – group by – apply summaries and filters if required and select a template from the available templates and click finish

Q6.   A Preview Picture does not appear when you select a template. How can you rectify this?

E. Verify the database connection

F. Press F5 within the available templates section

G. Open the report from the templates folder – select file, Summary Info and check the Save Preview Picture checkbox

H. Press F5 within windows explorer

Q7.   You would like to reapply a template which you previously removed. Which of the following methods will allow you to reapply the template?

A. Choose File | Template Expert | select template to be applied and click the 'reapply previous template' radio button and click OK

B. Choose View | Template Expert | select template to be applied and click the 'Re-Apply Last Template' radio button and click OK

C. Choose Format | Template Expert | select template to be applied and click the 'Re-Apply Last Template' radio button and click OK

D. Choose Report | Template Expert | and click the 'Re-Apply Last Template' radio button and click OK

Q8. **You want to remove the template, which you applied to your report. What should you do?**

A. Choose File | Template Expert | select template to be applied and click the 'Undo the current template' radio button and click OK
B. Choose Report | Template Expert | and click the 'Undo the current template' radio button and click OK
C. Choose View | Template Expert | select template to be applied and click the 'Undo the current template' radio button and click OK
D. Choose Format | Template Expert | select template to be applied and click the 'Undo the current template' radio button and click OK

Q9. **You want to undo the report template applied to a report; however the undo button is greyed out. What could be the possible reason? *(Multiple Answers)***

E. The report was initially created using a template, hence it cannot be removed
F. You have closed the report down and reopened it
G. You have not refreshed the report
H. None of the above

Q10. **You want to specify the type of data, which should appear within your template field. Which of the following must you implement?**

A. Right-Click the Template field and select edit template, replace the space(10) with the database field type
B. Right-Click the Template field and select edit template, replace the space(110) with the database field type
C. Right-Click the Template field and select edit template, replace the space(11) with the database field type
D. From the Field Explorer Right-Click the Template field and select edit template, replace the space(10) with the database field type

Q11. **Which of the following file extensions apply to a template?**

A. .xls
B. .xlst

C. .xml

D. .rpt

Q12. **A template field object can only be applied to which of the following?** *(Multiple Answers)*

A. Database Field

B. Formulas

C. Parameter Field

D. Special Field

E. SQL Expression Field

CHAPTER 3 - FORMULAS

Chapter 3 covers questions on formula creation within Crystal Reports and the syntax types, functions and operators, you will be tested on formula error detection and modification and using the Formula Workshop.

Keywords:
Formula, Parentheses, Formula Workshop, Syntax, Functions, Operators, Financial, String, Math, Date, String, SQL Expressions, Control Structures, Variables, Arrays

Create a Formula

Q1.   Which of the following languages can be used to create formulas in Crystal Reports? *(Multiple Answers)*

A. Crystal syntax
B. Visual Basic
C. Basic Syntax
D. PL SQL

Q2.   Detect the Parentheses Operator in the following formula?

WhilePrintingRecords;
NumberVar Total;
Total:= tonumber({Sales.Value})

A. :=
B. =
C. NumberVar
D. Total_Score

Q3.   Formulas cannot be used as groups

A. True
B. False

Q4. **You are creating a formula within the Formula Workshop - Formula Editor. Which of the following will produce a list of functions?**

A. Ctrl-space
B. Alt F9
C. Ctrl Alt F8
D. F8
E. F1

Q5. **Within which of the following sections can you rename a formula?** *(Multiple Answers)*

A. Function Workshop
B. Operators Workshop
C. Field Explorer - Formula Fields – Select Formula – Press (F2) to Rename or Right-click and select Rename from the drop-down menu
D. Within the Formula Workshop– Select Formula – Press (F2) to Rename or Right-click and select Rename from the drop-down menu

Q6. Which of the following statements are true? *(Multiple Answers)*

A. .(period) separates the table name from the field name
B. {} surrounds the database fields, formula names and parameter fields
C. ' 'separates the table name from the field name whilst using a database field
D. \\ Denotes a statement
E. ( ) Denotes a statement

Q7. **Which of the following apply to formula commenting?** *(Multiple Answers)*

A. Explains formula functionality
B. // Denotes a comment in Crystal
C. \\ Denotes a comment in Crystal
D. Can be used to deciphers the syntax used
E. Can be used to provide information for future programming changes

Q8. You would like to insert a new formula. What should you do?

A. Choose Report | Field Explorer | right click the formula fields and select New, enter formula name, this should bring you to the Formula Workshop -Formula Editor
B. Choose View | Field Explorer | right click the formula fields and select New, enter formula name, this should bring you to the Formula Workshop -Formula Editor
C. Choose Database | Field Explorer | right click the formula fields and select New, enter formula name, this should bring you to the Formula Workshop - Formula Editor
D. Choose Format | Field Explorer | right click the formula fields and select New, enter formula name, this should bring you to the Formula Workshop - Formula Editor

Q9. Which of the following is used to separate multiple arguments in functions?

A. ;
B. :
C. :=
D. ?
E. ,

Q10. What does the following formula represent when it appears in the report footer: - Sum({Sales.SalesAmount})

A. A sum of sales by Group Header
B. Grand Total for entire report
C. A sum of sales by Page Header
D. A sum of sales by Group Footer

Q11. Which of the following is used for Basic Syntax commenting?

A. // Denotes a comment in Basic Syntax
B. \\ Denotes a comment in Basic Syntax
C. 'Denotes a comment in Basic Syntax
D. . Denotes a comment in Basic Syntax
E. __Denotes a comment in Basic Syntax

Q12. Which of the following signs precedes a formula?

A. $
B. //
C. :
D. ?
E. @

Q13. Which of the following must be used to separate the statements in your formula?

WhilePrintingRecords
NumberVar Total_Score
Sales_Total:= tonumber({Sample_Exam_Results_2005.SCORE})
Sales_Total

A. //
B. :
C. :=
D. ?
E. ,
F. ;

Q14. Which of the following would you include when commenting your formulas? (Multiple Answers)

A. Brief explanation of the formula
B. Explanation of function
C. Explanation of database structure
D. Field used and why
E. Definition of variable

Q18. Which one of the following keys will activate HELP?

A. F2
B. F1
C. F3
D. F4

Q19. You want to duplicate a formula. Which of the following apply?

A. Right-click an existing formula within the Workbench and select duplicate
B. Right-click an existing formula within the Field Explorer and select duplicate
C. Right-click an existing formula within the Report Explorer and select duplicate
D. Right-click an existing formula within the Explorer and select duplicate

Q20. You want to delete a formula you have created; you cannot delete the formula because it is still being used within your report. How can you easily locate where the formula is being used? *(Multiple Answers)*

A. Right-click an existing formula within the Workbench and select Find in Formulas
B. Right-click an existing formula within the Report Explorer and select Find in Formulas
C. Right-click an existing formula within the Explorer and select Find in Formulas
D. Right-click an existing formula within the Field Explorer and select Find in Formulas

Q21. Using the Find in Formulas Function will perform a serach of all sections of the report including the subreport.

A. True
B. False

Use Functions and Operators - Financial

Q1. You have just sold your corporate bonds and would like to calculate the Accrued interest. Which of the following Crystal financial functions can be used to calculate the Accrued Interest for your bonds, which pays periodical interest?

A. AmorLINC
B. ACCRINT
C. AmorDEGRC
D. ACCRINTM
E. CoupNCD

Q2. You want to calculate the price per $200 face value of a security when the first period is odd. Which of the following functions can be used?

A. OddLYield
B. OddPrice
C. OddFPrice
D. OddFieldYield

Q3. Which of the following Crystal financial functions calculates the number of days from the settlement date to the next coupon date?

A. CoupDayBS
B. CoupDaysNC
C. CoupNCD
D. AmorDEGRC
E. ACCRINTM
F. CumlPRINC

Q4. Which of the following functions calculates the number of regular payments on a loan, or for an investment annuity?

A. NPER
B. XNPV
C. MIRR
D. CULM

E. IPMT
F. EFFEC
G. EFFECT

Q5. **Which of the following functions calculates the interest rate for a fully invested security?**

A. XIRR
B. XNPV
C. CUMIPMT
D. INTRATE
E. IPMT
F. EFFEC
G. EFFECT

Q6. **Which of the following Crystal financial functions can be used to calculate Accrued Interest for a bond, which pays interest on maturity?**

A. AmorLINC
B. ACCRINT
C. CoupNCD
D. AmorDEGRC
E. ACCRINTM

Q7. **Your client procured bonds with a maturity date in the next five years. Which of the following functions will compute the interest payments between a given settlement date and maturity date?**

A. COUPNUM
B. CoupDayBS
C. CoupDaysNC
D. CoupNCD
E. ACCRINTM
F. CumlPRINC

Q8. **Which of the following Crystal Financial Function can be used to calculate the depreciation of an asset?** *(Multiple Answers)*

A. AmorLINC

B. ACCRINT
C. AmorDEGRC
D. DISC
E. ACCRINTM
F. CoupNCD

Q9. **Which of the following calculates the price on a discounted security?**

A. PRICEMATV
B. PRICEMAT
C. PRICEDISC
D. PRICEMATT

Q10. **Which of the following Crystal Financial Function calculates the number of days from a coupon's last payment date, to the settlement date?**

A. CoupDayBS
B. ACCRINT
C. CoupNCD
D. AmorDEGRC
E. ACCRINTM
F. CumlPRINC

Q11. **Which of the following Crystal Financial Functions calculates the number of days in the coupon period, including the settlement date?**

A. CoupDayBS
B. CoupDays
C. CoupNCD
D. AmorDEGRC
E. ACCRINTM
F. CumlPRINC

Q12. **You want to calculate the interest rate applicable to a loan. Which of the following functions apply?**

A. RateDV
B. Rate

C. RateInt
D. Pricematv
E. Pricedisc

Q13. **You want to calculate the yield value of a security when the first period is odd. Which of the following functions can be used?**

A. OddPriceYield
B. OddFYield
C. OddLPrice
D. YieldPrice

Q14. **Which of the following will calculate the price value of a security, where interest is paid upon maturity?**

A. PRICEMATV
B. PRICEMAT
C. PRICEDISC
D. PRICEMATT

Q15. **Which of the following Crystal Financial Function calculates the future value of a loan?**

A. DISC
B. Ipmt
C. ISPMT
D. Nper
E. FV
F. CoupNCD

Q16. **Which of the following calculates the next coupon date following a given settlement date?**

A. DISC
B. Ipmt
C. ISPMT
D. CoupPCD
E. FV
F. CoupNCD

Q17. Which of the following calculates the internal rate of return?

A. IRR
B. IRTRN
C. IRRTRN
D. Nper
E. OddFYield

Q18. Your client has taken out a loan, and you want to calculate the cumulative interest paid. Which of the following apply?

A. DISC
B. DB
C. CUMPRINC
D. CUMIPMT
E. CULM
F. DDB

Q19. Your client has taken out a loan, and you want to calculate the cumulative principal paid. Which of the following apply?

A. DISC
B. DB
C. CUMPRINC
D. CUMIPMT
E. CULM
F. DDB

Q20. Which of the following converts a dollar decimal price to a fractional dollar price?

A. DollarFR
B. DollarFFR
C. DollarDDE
D. DollarDE

Q21. Given the interest rate, the term, present value, and future value. Which of the following calculates a payment?

A. PMT
B. PPMT
C. CUMPRINC
D. CUMIPMT
E. CULM
F. DDB

Q22. Which of the following calculates fixed declining balance depreciation, for an asset?

A. DISC
B. DB
C. CUMPRINC
D. CUMIPMT
E. CULM
F. DDB

Q23. Which function calculates double-declining balance depreciation, for an asset?

A. DISC
B. DB
C. CUMPRINC
D. CUMIPMT
E. CULM
F. DDB

Q24. Which of the following functions calculates the principal portion of a payment?

A. PPmt
B. PMMT
C. PV
D. MPT

Q25. Which function calculates the discount rate for a security?

A. DISC

B. DB
C. CUMPRINC
D. CUMIPMT
E. CULM
F. DDB

Q26. **Which function calculates the effective annual interest?**

A. DISC
B. EFF
C. CUMIPMT
D. CULM
E. DDB
F. EFFEC
G. EFFECT

Q27. **Which of the following functions is used to gauge the impact on a bonds price, when the bond yields changes?**

A. MyieldBond
B. MDuration
C. MIRR
D. NPer

Q28. **Which of the following functions calculates Macaulay duration?**

A. DUR
B. DURATION
C. DRR
D. DDDRR
E. DR

Q29. **Which of the following functions calculates the internal rate of return, for an investment?**

A. XIRR
B. EFF
C. CUMIPMT
D. CULM
E. DDB

F. EFFEC

G. EFFECT

Q30. Which of the following functions calculates the net present value, for an investment?

A. XIRR

B. XNPV

C. CUMIPMT

D. CULM

E. DDB

F. EFFEC

G. EFFECT

Q31. Which of the following functions calculates the present or future value of an annuity?

A. PVV

B. PPV

C. PPBB

D. PV

Q32. Which of the following functions calculates the interest portion of a payment?

A. XIRR

B. XNPV

C. CUMIPMT

D. CULM

E. IPMT

F. EFFEC

G. EFFECT

Q33. Which of the following functions calculates the nominal annual interest?

A. CoupDayBS

B. ISPMT

C. CoupNCD

D. Nominal

E. ACCRINTM

F. CumlPRINC

Q34. Which of the following functions calculates the modified internal rate of return implicit in a set of cash flows?

A. XIRR
B. XNPV
C. MIRR
D. CULM
E. IPMT
F. EFFEC
G. EFFECT

Q35. Which of the following functions calculates the straight-line interest portion of a payment?

A. ISPMT
B. XNPV
C. MIRR
D. CULM
E. IPMT
F. EFFEC
G. EFFECT

Q36. Which of the following functions calculates straight-line balance depreciation, for an asset?

A. ISPMT
B. SLN
C. MIRR
D. CULM
E. IPMT
F. EFFEC
G. EFFECT

Q37. Which of the following functions calculates the sum of the year's digits depreciation, for an asset?

A. SYD
B. SLN
C. VDB
D. CULM

E. IPMT

F. EFFEC

G. EFFECT

Q38. **Which of the following functions calculates declining balance depreciation, for an asset?**

A. SYD

B. SLN

C. VDB

D. CULM

E. IPMT

F. EFFEC

G. EFFECT

Q39. **Which of the following functions calculates a payment?**

A. Pmt

B. PMMT

C. PV

D. MPT

Q40. **Which of the following functions calculates the interest rate implied to a mortgage?**

A. Rate

B. Yield

C. PV

D. Pmt

Q41. **Which of the following functions calculates the future value of a completely invested coupon?**

A. ISPMT

B. XNPV

C. MIRR

D. CULM

E. RECEIVED

F. EFFEC

G. EFFECT

Q42. Which of the following financial functions calculates the yield of a security, a discounted yield, and the yield of a security where interest is paid on maturity? *(Multiple Answers)*

A. OddPriceYield
B. OddFYield
C. YieldDisc
D. YieldMat
E. YieldPrice
F. Yield

Q43. Which of the following financial functions are utilized for Treasury bill calculations? *(Multiple Answers)*

A. TBILLEQ
B. TBILLPRICE
C. TBILLYIELD
D. TBILLYLD
E. TBILLPRC

Functions and Operators – Date Functions and Formulas

Q1. Convert 4/10/2008 to a date time format. Which script applies? 4/10/2008

- A. #4/10/2008#
- B. #4/10/2008
- C. 4/10/2008##
- D. ##4/10/2008##
- E. CDateTime(4/10/2008)
- F. CurrentDateTime(4/10/2008)

Q2. You want to extract the month from the exam date. Which of the following apply?

- A. DatePart ('mm',{Exams.ExamsDate} ,1)
- B. DatePart ({Exams.ExamsDate},'m')
- C. DatePart ({Exams.ExamsDate},'mm')
- D. DatePart ('m', {Exams.ExamsDate})

Q3. Which of the following will display the Current Date and Time? *(Multiple Answers)*

- A. CurrentDate + "" + CurrentTime
- B. PrintDate + "" + PrintTime
- C. CurrentDateTime
- D. CurrentDate + CurrentTime
- E. PrintDate + PrintTime

Q4. Which of the following can you specify within a DateDiff Formula? *(Multiple Answers)*

- A. IntervalType
- B. Type
- C. LastDayofWeek
- D. BeginDate for Interval Calculation
- E. EndDate for Interval Calculation

Q5. **What does the following formula represent?**
Sum({Sales.Amount},{Sales.Date}, "weekly")

A. A sum of sales by Sales Rep Group Header
B. Sales amount subtotal ordered by the sales date by week
C. A sum of sales by Group Footer
D. Grand Total for entire report

Q6. **What will the following return?**
#09:09#

A. 9.9.00am
B. 9.9am
C. 09:09:00am
D. 9;9am

Q7. **You want to highlight all sales made on Saturday as 'Last Day of Week Sales' and all others as 'Others'. Using the Sales Date which of one the following apply?**

A. if DayOfWeek ({Orders.SalesDate}) = 'Saturday' then 'Last Day of Week Sales' else 'Others'
B. if WeekdayName({Orders.SalesDate}) = 'Saturday' then 'Last Day of Week Sales' else 'Others'
C. if WeekdayName (DayOfWeek ({Orders.SalesDate})) = 'Saturday' then 'Last Day of Week Sales' else 'Others'
D. if WeekdayName (DayOfWeek ({Orders.SalesDate})) = crSaturday then 'Last Day of Week Sales' else 'Others'

Q8. **What will the following formula return Date({Order. Ship}) – Date({Order. OrderDate})?**

A. Number of days between Dates
B. Converts datetime to date
C. Date
D. String

Q9. **How would you convert a string field type which contains date information?**

A. DateValue

B. Cdate
C. CurDate
D. Dval

Q10. **You want to create a formula to determine how many days are required before an order is due. Which of the following apply?**

A. "Order Number " + text({Orders.Order ID}) + " is required in " + text({Orders.Required Date} - {Orders.Order Date}) + "days"

B. "Order Number " + totext({Orders.Order ID}) + " is required in " + totext({Orders.Required Date} - {Orders.Order Date}) "days"

C. "Order Number " + tonumber({Orders.Order ID}) + " is required in " + todate({Orders.Required Date} - {Orders.Order Date}) + "days"

D. "Order Number " + totext({Orders.Order ID}) + " is required in " + totext(round(tonumber({Orders.Required Date} - {Orders.Order Date})),0) + "days"

Q11. **The rules of the competition state all customers who purchased a product on Tuesday will be eligible for a 15% discount on their next purchase. Which of the following apply?**

A. if day({Orders.OrderDate}) = crTue then 'Customer is eligible for 15% discount on next Purchase'

B. if day({Orders.OrderDate}) = crTuesday then 'Customer is eligible for 15% discount on next Purchase'

C. if day({Orders.OrderDate}) = Tuesday then 'Customer is eligible for 15% discount on next Purchase'

D. if day({Orders.OrderDate}) = Tue then 'Customer is eligible for 15% discount on next Purchase'

Q12. **The date and time are held in different fields, you want to display them together. Which of the following apply?**

A. DateTimeTime({Date},{Time})
B. TimeDate({Date},{Time})
C. DateTimeValue ({Date},{Time})
D. DateTime ({Date},{Time})

Q13. **What does the DateDiff function return?**

A. Number of data dates between dates
B. Number of digits between dates
C. Extracts number of Days, Months, Weeks, Years, Hours, Seconds between two specified dates
D. Converts Date to Datetime

Q14. **Which of the following are intervals that can be used within the DateDiff function?** *(Multiple Answers)*

A. yyyy
B. q
C. m
D. y
E. d
F. w
G. ww
H. h
I. n
J. s

Q15. **You have received a request to produce a report which highlights the previous month deliveries; (for this example the current month is December) these are deliveries made in November of the present year, and those delivered in the past seven days. If none of these deliveries fall within this category then display the delivery date only. Which of the following formulas apply?** *(Multiple Answers)*

A. if ({Delivery.DeliveryDate})in LastFullMonth then 'Last Months Orders' else if Month({Delivery.DeliveryDate})= 11 and Year({Delivery.DeliveryDate})= year(CurrentDate) then 'Current Year November Orders' else if {Orders.ShippedDate} in Last7Days then 'Orders Made in the Last Seven Days' else Totext({Orders.ShippedDate})

B. if month({Delivery.DeliveryDate})= Month(CurrentDate) - 1 then 'Last Months Orders' else if Month({Delivery.DeliveryDate})= 11 and Year({Delivery.DeliveryDate})= year(CurrentDate) then 'Current Year November Orders' else if {Orders.ShippedDate}

= Last7Days then 'Orders Made in the Last Seven Days' else Totext({Orders.ShippedDate})

C. if ({Delivery.DeliveryDate})in LastFullMonth then 'Last Months Orders' else if Month({Delivery.DeliveryDate})= 11 and Year({Delivery.DeliveryDate})= (CurrentDate) then 'Current Year November Orders' else if {Orders.ShippedDate} in Last7Days then 'Orders Made in the Last Seven Days' else ({Orders. ShippedDate})

D. if ({Delivery.DeliveryDate}) = LastFullMonth then 'Last Months Orders' else if Month({Delivery.DeliveryDate})= 11 and Year({Delivery.DeliveryDate})= year(CurrentDate) then 'Current Year November Orders' else if {Orders.ShippedDate} = Last7Days then 'Orders Made in the Last Seven Days' else Totext({Orders. ShippedDate})

Q16. You want to find the number of days between the Request Date and the Ship Date. Which of the following formulas apply?

A. dateserial("d", {Order. RequestDate}, {Order. ShipDate})
B. datepart("d", {Order. RequestDate}, {Order. ShipDate})
C. datemart("d", {Order. RequestDate}, {Order. ShipDate})
D. datediff("d", {Order. RequestDate}, {Order. ShipDate})

Q17. Which of the following formulas will produce 12 months rolling?

A. Dateadd("yyyy",-1, CurrentDate) to CurrentDate
B. Dateadd("y",-1, CurrentDate) to CurrentDate
C. Dateadd("yyyy",-2, CurrentDate) to CurrentDate
D. Dateadd("yyyy",-12, CurrentDate) to CurrentDate

Q18. Which formula will extract the dates from the Date field which also contains non-date fields?

A. If notisdate({DateField}) then ({DateField}) else Date(0,0,0)
B. If Isnumeric({DateField}) then ({DateField}) else Date(0,0,0)
C. If date({DateField}) then ({DateField}) else Date(0,0,0)
D. If Isdate({DateField}) then ({DateField}) else Date(0,0,0)
E. None of the above

Functions and Operators – Maths Functions and Formulas

Q1. Random numbers can be extracted using which of the following?

A. Sgn
B. Abs()
C. Random
D. Exp
E. Rnd
F. Truncate

Q2. Which of the following will produce 70 when applied to a result field containing 69.30?

A. Round({result},0)
B. RoundUp({result},0)
C. Rnd
D. Rnd({result},0)

Q3. You multiple a number field by a currency field. What will the result be?

A. Number field
B. Text Field
C. Currency Field
D. An error message will ensue

Q4. You receive an error message, which reads "The remaining text does not appear to be part of the formula" Stock. UnitCost} * { Stock.Quantity}. Which of the following will rectify this error?

A. Stock.UnitCost * { Stock.Quantity}
B. Stock.UnitCost} * { Stock.Quantity}
C. {Stock.UnitCost} * Stock.Quantity}
D. {Stock.UnitCost} * { Stock.Quantity}

Q5.   Which of the following will produce 64?

    A. 4^3 (Exponentiation)
    B. Sgn(3 to 4)
    C. Abs(3)~3
    D. Mod(3,3,2)

Q6.   Which of the following will produce 20.40 when applied to a result field containing 20.45?

    A. Truncate({result},1)
    B. Round({result},0)
    C. RoundUp({result},1)
    D. Rnd
    E. Rnd({result},0)

Q7.   You have created the following formula.
{Cost.TotalCost} / {Cost.UnitCost}

You receive the following error message :- 'Division by zero'
What should you do to rectify the problem?

    A. Multiple the {Cost.TotalCost} * {Cost.UnitCost}
    B. Multiple the {Cost.TotalCost} * {Cost.UnitCost} * -1
    C. If {Cost.TotalCost} = 0 then 0 else {Cost.TotalCost} / {Cost.UnitCost}
    D. If {Cost.UnitCost} = 0 then 0 else {Cost.TotalCost} / {Cost.UnitCost}

Q8.   Which of the following will produce the total amount of sales per representative?

    A. Sum(Sales.SaleRep},{Sales.sales.Amount})
    B. Sum({Sales.SaleRep},{Sales.sales.Amount})
    C. Sum({Sales.sales.Amount},{Sales.SaleRep})
    D. Sum({Sales.sales.Amount} & + "" +&{Sales.SaleRep})

Q9.   You have created the following formula. You receive the error message 'Division by zero'. What should you do to rectify the problem?

If {Client.ClientID} = "DRP" Then {Cost.TotalCost} / {Cost.UnitCost}

A. Multiple the {Cost.TotalCost} * {Cost.UnitCost}
B. Multiple the {Cost.TotalCost} * {Cost.UnitCost} * -1
C. If {Cost.UnitCost} = 0 then 0 else {Cost.TotalCost} / {Cost. UnitCost}

Q10. You have created the following formula and receive the error message 'The remaining text does not appear to be part of the formula'. How can you rectify this problem?

NumberVar TotalOrdSum: =Count ({Command.Client Order Amount})
If {Command.Client ID} = 501 Then
TotalOrdSum: = Count ({Command.Client Order Amount})
else 0

A. NumberVar TotalOrdSum: =Count ({Command.Client Order Amount}) If {Command.Client ID} = 501 Then TotalOrdSum: = Count ({Command.Client Order Amount}) 0
B. NumberVar TotalOrdSum:=Count({Command.Client Order Amount}) if {Command.Client ID} = 501 Then TotalOrdSum:= Count({Command.Client Order Amount});
C. NumberVar TotalOrdSum:= Count({Command.Client Order Amount}); if {Command.Client ID} = 501 Then TotalOrdSum:= Count({Command.Client Order Amount}) else 0
D. NumberVar TotalOrdSum:=Count({Command.Client Order Amount}) if {Command.Client ID} = 501 Then TotalOrdSum:= Count({Command.Client Order Amount}) Else NOT 501

Q11. Which of the following will convert (888.300) to 888?

A. Sgn
B. Abs()
C. Int
D. Truncate

Q12. You want to convert a value to a currency. Which of the following apply? (Multiple Answers)

A. ${Table.Value}

B. CCur({Table.Value})
C. CStr({Table.Value})
D. CDbl({Table.Value})
E. DB({Table.Value})

Q13. Which of the following Crystal functions will give you 503? When applied to: 903 – 1406

A. PopulationStdDev()
B. StdDev
C. Variance
D. MakeArray

Q14. Which of the following converts a fractional dollar price to a decimal price?

A. DollarFR
B. DollarFFR
C. DollarDDE
D. DollarDE

Q15. You apply a summary (Sum Grand Total to the Total figures below). Which of the following results will be produced?

Record Number	Total
1	0
2	6
3	10
4	0
5	0
6	0
7	9
8	0
9	0
10	1
11	25
12	60

A. 310
B. 119
C. 112
D. 111

Q16. **You apply a summary (Distinct Count to the data below). Which of the following results will be produced?**

Record Number	Total
1	1
2	1
3	1
4	0
5	0
6	0
7	0
8	0
9	0
10	0
11	0
12	0

A. 3
B. 9
C. 12
D. 2
E. 10

Q17. You apply a summary (Count to the data below). Which of the following results will be produced?

Record Number	Total
1	1
2	1
3	1
4	0
5	0
6	0
7	0
8	0
9	0
10	0
11	0
12	0

A. 3
B. 9
C. 12
D. 2
E. 10

Q18. A request has been made to highlight the discounts that will be assigned to clients based on their orders. For orders less than or equal to 40, a 5% discount will be awarded, between 41 and 80, 10% and 81 and over, 15%. Which of the following formulas apply?

A. A.   if count({Order_Details.Quantity}) = 40 then 'Eligible for 5% Discount' else if count({Order_Details.Quantity}) <= 41 and count({Order_Details.Quantity})<= 80 then 'Eligible for 10% Discount' else if count({Order_Details.Quantity}) = 81 then 'Eligible for 15% Discount'

B. B.   if count({Order_Details.Quantity}) <= 41 then 'Eligible for 5% Discount' Else if count({Order_Details.Quantity}) >= 41 and count({Order_Details.Quantity})<= 80 then 'Eligible for

10% Discount' else if count({Order_Details.Quantity}) >= 81 then 'Eligible for 15% Discount'

C. C.   if count({Order_Details.Quantity}) >= 40 then 'Eligible for 5% Discount' Else if count({Order_Details.Quantity}) >= 41 and count({Order_Details.Quantity})<= 80 then 'Eligible for 10% Discount' Else if count({Order_Details.Quantity}) >= 81 then 'Eligible for 15% Discount'

D. D.   if count({Order_Details.Quantity}) <= 40 then 'Eligible for 5% Discount' Else if count({Order_Details.Quantity}) >= 41 and count({Order_Details.Quantity})<= 80 then 'Eligible for 10% Discount' Else if count({Order_Details.Quantity}) >= 81 then 'Eligible for 15% Discount'

Q19.   You want to extract only fields which contain numbers from the delivery day's field.   Which of the following apply? *(Multiple Answers)*

A. If Numeric({Ship.deliverydays}) then {Ship.deliverydays})
B. If ToNumber({Ship.deliverydays}) then {Ship.deliverydays})
C. If IsNumeric({Ship.deliverydays}) then {Ship.deliverydays})
D. If IsNumber({Ship.deliverydays}) then {Ship.deliverydays})
E. If NumericText({Ship.deliverydays}) then {Ship.deliverydays})

Q20.   You apply Floor to the following value £3,479.70.   What will it produce?

A. 3470
B. 3479
C. 3460
D. 3480

Q21.   You apply Round to the following value £3,479.70.   What will it produce?

A. 3470
B. 3479
C. 3460
D. 3480

Q22. You apply Ceiling to the following value £3,479.70. What will it produce?

A. 3470
B. 3479
C. 3460
D. 3480

Functions and Operators – String
Functions and Formulas

Q1. How would you concatenate the following string fields?

Author.Title
Author.FName
Author.LName

    A. Author.Title & "Author.FName" & "Author.LName"
    B. Author.Title & " " & Author.FName & " " & Author.Lname
    C. "Author.Title" & "Author.FName" & "Author.LName"
    D. "Author.Title" *"Author.FName" *"Author.LName"

Q2. You want to extract the first letter of the Authors first name to produce the following: - Ms A.Iroko, using the following fields

Author.Title, Author.FName, Author.LName

    A. Author.Title & " " & Author.Fname =[1] & "." Author.Lname
    B. Author.Title & " " & Author.Fname=1 & "." Author.Lname
    C. Author.Title & " " & Author.Fname .1 & "." Author.LName
    D. Author.Title & " " & Author.Fname [1] & "." Author.Lname
    E. Author.Title & " " & Author.Fname [1] & "."& Author.LName

Q3. Which of the following is a Subscript formula?

    A. {Author.FullName} [5 to 8]
    B. {Author.FullName} (5 to 8)
    C. {Author.FullName} {5 to 8}
    D. {Author.FullName} '5 to 8'

Q4. You want to display 1470 as one thousand four hundred and seventy. Which of the following apply?

    A. ToNumber(1470,0)
    B. ToWords(1470,0)
    C. ToNumbers(1470,0)
    D. ToWord(1470,0)
    E. ToText(1470,0)

Q5. You want to display the Firstname field on your report in lowercase letters, however the First letter of the Firstname must be displayed as a capital letter. Which of the following will achieve this aim?

A. ProperCase({FirstName})
B. lowercase({FirstName})
C. lowercase(uppercase({FirstName}))
D. Proper({FirstName})

Q6. The message below is displayed when you create the formula: Which of the following will rectify the problem?

"The remaining text does not appear to be part of the formula".

CurrencyVar SaleValue;
StringVar CustName
DateTimeVar ShipDate;

A. CurrencyVar SaleValue StringVar CustName; DateTimeVar ShipDate;
B. CurrencyVar SaleValue StringVar CustName DateTimeVar ShipDate
C. CurrencyVar SaleValue; StringVar CustName; DateTimeVar ShipDate ;
D. CurrencyVar SaleValue: StringVar CustName; DateTimeVar ShipDate ;

Q7. How can you eliminate spaces between text in your database table?

A. Truncate({Customer.Name})
B. Truncate({Customer.Name},1)
C. Trun({Customer.Name})
D. Trim({Customer.Name})

Q8.   You created a formula as follows

" Thank You for purchasing £ " + {Customer.Sales} + " in goods last year."
You receive the following error message : 'A string is required here'. What should you do to rectify the problem?

    A. " Thank You for buying £ " + ({Customer.Sales})+ " in goods last year."

    B. " Thank You for buying £ " + Customer.Sales + " in goods last year."

    C. " Thank You for buying £ " + {Customer.Sales} + " in goods last year.

    D. Thank You for buying £ " + {Customer.Sales} + " in goods last year."

    E. " Thank You for buying £ " + ToText({Customer.Sales}) + " in goods last year."

Q9.   If a CustomerID begins with LON then display' London based customers', otherwise display 'other cities'. Which of the following formulas will produce the result required? *(Multiple Answers)*

    A. if {Orders.CustomerID} startswith "LON" then "London based clients" else "other cities"

    B. if {Orders.CustomerID} like "LON" then "London based clients" else "other cities"

    C. if {Orders.CustomerID} is "LON" then "London based clients" else "other cities"

    D. if {Orders.CustomerID} is like %LON then "London based clients" else "other cities"

    E. IIF ({Orders.CustomerID} Startswith 'LON', ' London based clients,' other cities ')

    F. Switch ({Orders.CustomerID} Startswith 'LON', "London based clients", NOT ({Orders.CustomerID}Startswith 'LON'), " other cities ")

Q10.   The title, first name and last name, are held in different fields. You want to display all three in the same formula, putting a dot between the title and first name, last name, and you also want the city to appear in brackets beside

the last name and in uppercase. Which of the following formulas will achieve this? *(Multiple Answers)*

Example: Ms.Antonia.Iroko (LONDON)

A. {Clients.Title} + "." + {Clients.FirstName} + " " + {Clients.LastName} + " (" + uppercase({Clients.City})+ ") "
B. {Clients.Title} + "." + {Clients.FirstName} + " " + {Clients.LastName} + " (" + upcase({Clients.City})+ ") "
C. {Clients.Title} + "." + {Clients.FirstName} + " " + {Clients.LastName} + " (" + uppercase({Clients.City})+ ") " +
D. {Clients.Title} & "." & {Clients.FirstName} & "." & {Clients.LastName} & " (" & uppercase({Clients.City}) & ") "
E. {Clients.Title} & "." & {Clients.FirstName} & "." & {Clients.LastName} & " (" & ucase({Clients.City}) & ") "

Q11.   Which of the following is a Boolean formula?

A. if ({customer.region}) = 'al' then true else false
B. If {@Submit hours} = 0 then "Submitted on Time" else If {@Submit hours} = 1 then "Submitted in 1 Hour" else "Submitted in " + ToText({@Submit hours}, 0) + "hours"
C. ToText ({@1000s formula},2,',',':')
D. Create the following formula: distinctCount({Client.ClientID},{Client.Country}) and place it in the details section

Q12.   Which of the following will convert a datetime datatype to a string? *(Multiple Answers)*

A. Cstr()
B. ToText()
C. ToWord()
D. C()

Q13.   What will be the result of the following formula?

"201" & "301" & "401"

A. 75
B. Syntax Not recognized
C. 201301401

Q14. **CStr applied to 19.99 will return which of the following?**

    A. Nineteen ninety nine and ninety nine
    B. 19.99
    C. Converts 19.99 to a string
    D. 20

Q15. **You have been asked to change the following formula to a CASE statement. Which of the following is correct?**

if {Ship.ShipDays} = 0 then "Shipped Same Day" else if {Ship.ShipDays} = 1 then "Shipped in 1 Day" else "Shipped in " + ToText({Ship.ShipDays}, 0) + "Days"

    A. Select {Ship.ShipDays} CASE 0: "Shipped On Time" CASE 1: "Shipped in 1 Day" Default: "Shipped in " + ToText({Ship.ShipDays} , 0) + "Days"
    B. Select {Ship.ShipDays} CASE: "Shipped On Time" CASE : "Shipped in 1 Day" Default: "Shipped in " + ToText({Ship.ShipDays} , 0) + "Days"
    C. Select {Ship.ShipDays} CASE 0 "Shipped On Time" CASE 1 "Shipped in 1 Day" Default: "Shipped in " + ToText({Ship.ShipDays} , 0) + "Days"
    D. Select {Ship.ShipDays} CASE 0: "Shipped On Time" CASE 1: "Shipped in 1 Day" Default: "Shipped in " + ({Ship.ShipDays} , 0) + "Days"

Q16. **Within the Formula Workshop - Formula Editor. Which of the following can be accessed?** *(Multiple Answers)*

    A. Functions
    B. Operators
    C. Reports Fields
    D. Database Tables and Fields

Q17. You want to present a report which displays the maximum number of each product purchased, and the Customer who purchased the product, your report contains a group based on product name. Which of the following formulas will give you the results required?

A. 'The Maximum quantity of' + " " + ({Products.Name}) + " " + 'ordered by' + " " + {Customers.ID} + " " + 'is' + " " + totext(Maxi ({Order_Details.Quantity}, {Products.Name}))

B. 'The Maximum quantity of' + " " + ({Products.Name}) + " " + 'ordered by' + " " + {Customers.ID} + " " + 'is' + " " + totext(Maximum ({Order_Details.Quantity}, {Products.Name}))

C. 'The Maximum quantity of' + " " + ({Products.Name}) + " " + 'ordered by' + " " + {Customers.ID} + " " + 'is' + " " + totext(Max ({Order_Details.Quantity}, {Products.Name}))

D. 'The Maximum quantity of' + " " + ({Products.Name}) + " " + 'ordered by' + " " + {Customers.ID} + " " + 'is' + " " + Maximum ({Order_Details.Quantity}, {Products.Name})

Q18. Which function has been applied to the 'firstname' string that will produce the results below?

NAME	RESULT
Robert	65.00
Rita	78.00
Harry	76.00
Matthew	74.00

A. StrCmp()
B. Filter()
C. AscW()
D. Ascci()
E. Space()

Q19. Which functions has been applied to the Record No. to produce the result?

Record No	Result
1	I
2	11
3	111
6	VI

A. Convert()
B. Tonumber()
C. Roman()
D. Totext()

Q20. You want to extract the first four letters of the region field. Which of the following formulas will produce the correct result? *(Multiple Answers)*

A. {Customer. Region}[4]
B. {Customer. Region}[1 TO 4]
C. {Customer. Region}[1&2&3&4]
D. Left ({Customer. Region}, 1,4)
E. Left ({Customer. Region}, 4)
F. Mid ({Customer. Region}, 1,4)
G. TrimLeft ({Customer. Region}, 1,4)

Q21. Your formula produces the following error message "a string is required here". How do you rectify this error?

if {@Submit days} = 0 then "Submitted on Time" else {@Submit days}

A. if ({@Submit days}) = 0 then "Submitted on Time" else ({@ Submit days})
B. if (({@Submit days}) = 0 then "Submitted on Time" else ({@ Submit days}))
C. if ToText ({@Submit days}) = 0 then "Submitted on Time" else ({@Submit days})
D. if ({@Submit days}) = 0 then "Submitted on Time" else ToText({@ Submit days} , 0)

Q22. A number has been entered between characters in the lastname field. Which of the following formulas will highlight the data error entry?

A. if Roman({Customer.Lastname}) then 'Data Input Error'else {Customer.Lastname}
B. if not NumericText ({Customer.Lastname}) then 'Data Input Error'else {Customer.Lastname}
C. if ProperCase ({Customer.Lastname}) then 'Data Input Error'else {Customer.Lastname}

D. if Val({Customer.Lastname}) then 'Data Input Error' else {Customer.Lastname}

Q23. **Old student registration codes started with a prefix of their country, e.g. UK users will have a code of UK876655, and US users with a prefix of US786666. The length of the string old code was eight and has now changed to six. New codes will only contain the digits. You now want to display only digits without the prefix?** *(Multiple Answers)*

A. if length({@reg_code})= 8 then mid({@reg_code},3,6) else {@reg_code}

B. right({@reg_code},6)

C. left({@reg_code},6)

D. mid({@reg_code},4,7)

Use Functions and Operators – SQL Expressions

Q1. Which of the following methods can be used to access a SQL Expression field?

A. Choose View | Field Explorer | Right-Click the SQL Expression fields and click new.

B. Choose file | Field Explorer | Right-Click the SQL Expression fields and click new.

C. Choose View | Report |Field Explorer | Right-Click the SQL Expression fields and click new.

D. Choose Database| Report |Field Explorer | Right-Click the SQL Expression fields and click new.

Q2. Which SQL Expression will display the Database name?

A. UserDB

B. Database()

C. Database{}

D. Database[]

Q3. Which of the following will display the database user?

A. User()

B. UserDB

C. Database{}

D. Database[]

Q4. Which of the following are SQL Expression Functions? *(Multiple Answers)*

A. REPEAT(,)

B. BIT_LENGTH()

C. REPLACE(,,)

D. SOUNDEX()

E. OCTET_LENGTH()

Q5. **Which of the following is used to depict an SQL Expression name?**

A. @
B. ~
C. #
D. ?
E. %

Q6. **SQL Functions change in the SQL Expression Editor, depending on the type of database being used.**

A. True
B. False

Q7. **You have placed a SQL Expression field into your SQL language. What will appear?**

A. SQL Expression Name
B. SQL Expression formula
C. It will not show in the SQL Query
D. Error message

Q8. **You can use SQL expression fields within SQL Commands and Stored Procedures.**

A. True
B. False

Q9. **SQL Expressions improve performance of reports.**

A. True
B. False

Q10. **What are the basic requirements for using SQL Expressions? *(Multiple Answers)***

A. SQL Database
B. ODBC Database
C. .dll extension file
D. .ora drivers required

Use Control Structures

Q1. **Which of the following best describes a Control Structure?** *(Multiple Answers)*

A. A formula
B. A process
C. An element procedure
D. A statement which determines the order of execution of the other statements
E. The use of a formula to evaluate a process until a condition is encounter

Q2. **Which of the following can be defined as a control structure?** *(Multiple Answers)*

A. if WeekdayName (DayOfWeek ({Orders.SalesDate})) = 'Wednesday' then 'Required' else 'Not Required'
B. if ({Delivery.DeliveryDate})in LastFullMonth then 'Last Months Orders' else if Month({Delivery.DeliveryDate})= 11 and Year({Delivery.DeliveryDate})= year(CurrentDate) then 'Current Year November Orders' else if {Orders.ShippedDate} in Last7Days then 'Orders Made in the Last Seven Days' else Totext({Orders.ShippedDate})
C. ProperCase({Surname})
D. Select {Dispatch.DispatchDays} CASE 0: "Dispatched On Time" CASE 1: "Dispatched in 1 Day" Default: "Dispatched in " + ToText({Dispatch.DispatchDays} , 0) + "Days"

Q3. **Which of the following statements is true in relation to a 'For Loop'?**

A. The For Loop, is a counter variable
B. The For Loop, keeps track of how many times a variable is used in a program
C. The For Loop, is a specific piece of logic has been cycled though
D. The For Loop, uses a counter variable to determine how many times a particular part of logic has been rotated though

Q4. **Which of these statements is true in relation to a 'While Do Loop'?**

A. The While Do Loop evaluates a condition after each occurrence of the loop and stops if the condition is no longer true

B. The While Do Loop evaluates a condition during each occurrence of the loop and stops if the condition is no longer true

C. The While Do Loop halts when a condition defined is no longer true

D. The While Do Loop evaluates a condition and continues if the condition is no longer true

Q5. **Which of the following statements apply to an 'Option Loop'? *(Multiple Answers)***

A. Denote the maximum number of loops

B. Used for loop counts over 100000

C. Used for loop counts over 10000

D. Used for loop counts over 10000000

Q6. **Which of the following are control structures? *(Multiple Answers)***

A. do while

B. exit for

C. exit while

D. for := to step do

E. if then else

F. option loop

G. select case : default:

H. while do

I. while

Q7. **Which of the following apply to a 'While Do' loop?**

A. Assesses the condition and if the condition is reached it stops

B. Assesses the condition for a specified number only

C. Assesses the condition and if the condition is achieved it performs the application of the condition set in the do section of the code, the procedure is continued until the condition is no longer applicable

D. Loops through the value of a specified number of times?

E. None of the above

Q8.   Which of the following apply to a 'For Loop'?

A. Loops through the values of a specified period?

B. Assesses the condition and if the condition is reached it stops

C. Assesses the condition for a specified number only

D. Assesses the condition and if the condition is achieved it performs the application of the condition set in the do section of the code, the procedure is continued until the condition is no longer applicable

Use Variables

Which of the following statements are true? *(Multiple Answers)*

A. A local variable is not visible in a Subreport
B. A local variable in one formula cannot be accessed in another formula
C. You can access the value of a local variable in one formula from a different formula
D. You can share a local variable with a Subreport
E. You can access the value of a Global Variable in one formula from a different formula
F. A Global Variable declared in one formula can be accessed by another formula
G. A Global Variable cannot be seen in a Subreport
H. You can pass a Shared variable from a Subreport to the Main report
I. You can pass a Shared variable from a from a Main report to a Subreport

Q2. **Which of the following can be declared as variables within Crystal Reports?** *(Multiple Answers)*

A. CurrencyVar
B. TimeVar
C. TextVar
D. DateTimeVar
E. NumberVar
F. BooleanVar
G. StringVar
H. DateVar
I. DataVar

Q3. **Which of the following is a default variable?**

A. Local variable
B. Global variable
C. Shared variable
D. Variable

Q4. **Which of the following variable can be shared with Subreports and a main report?**

A. Local variable
B. Global variable
C. Shared variable
D. Variable

Q5. **Which of the following variables is being defined?**

'Can only be used within the formula it is declared in'

A. Local variable
B. Global variable
C. Shared variable
D. Variable

Q6. **Which of the following have to be done when creating a variable?**

A. Declare
B. Assign
C. Reference
D. Edit

Use Arrays

Q1. An additional certification exam code (RDCR08500a) has been assigned to the current exam certification path. You want to add the new code to your existing codes. Which of the following methods will allow you to add the new code and display it in your report?

    A. A.   Local StringVar array ExamCodes := ["RDCR08200a", "RDCR08300a", "RDCR08400a"];
Redim ExamCodes [4]; ExamCodes [4] := "RDCR08500a";
join(ExamCodes,",")

    B. B.   Local StringVar array ExamCodes := ["RDCR08200a", "RDCR08300a", "RDCR08400a"];

    C. Redim Preserve ExamCodes [4];ExamCodes [4] := "RDCR0850 0a";join(ExamCodes,",")

    D. C.   Local StringVar array ExamCodes := ["RDCR08200a", "RDCR08300a", "RDCR08400a"]; Preserve ExamCodes [4]; ExamCodes [4] := "RDCR08500a";

    E. join(ExamCodes,",")

    F. D.   Local StringVar array ExamCodes := ["RDCR08200a", "RDCR08300a", "RDCR08400a"]; Redim ExamCodes (4); ExamCodes (4) := "RDCR08500a";

    G. join(ExamCodes,",")

    H. E.   Local StringVar array ExamCodes := ["RDCR08200a", "RDCR08300a", "RDCR08400a"]; Redim ExamCodes (4); ExamCodes (4) := "RDCR08500a";
join(ExamCodes,",")

Q2. Arrays cannot be shared between a main report and a Subreport.

    A. True
    B. False

Q3. Which of the following is the Redim syntax in Crystal?

    A. Redim X[n]
    B. Redim x( )
    C. Redim x { }
    D. Redim x | |

Q4.   Which of the following is the Redim syntax in Basic?

A. Redim X[n]
B. Redim x( )
C. Redim x { }
D. Redim x | |

Q5.   What will the following function produce?

[({Sales.Amount *.6) , ({Sales.Amount *.6) , ({Sales.Amount *.75)]

A. Subscript
B. In Array
C. Make Array
D. Array
E. An error message

Q6.   What will the following array return?

Sum([95,35,10,2])

A. 142
B. 100
C. 4
D. 0
E. NONE OF THE ABOVE

Q7.   Which one of the following statements is true? *(Multiple Answers)*

A. Array subscripts start at 1 in Crystal.
B. Array subscripts start with 0, (default in VB).
C. Crystal does not support arrays with zero elements.
D. Crystal does not support arrays with 1 element.

Q8.   Which of the following are not variable declarations in Crystal? *(Multiple Answers)*

A. BooleanVar

B. NumberVar
C. CurrencyVar
D. DateVar
E. RangeVar
F. ArrayVar

Q9.   What are the data types in an array described as?

A. Array functionalities
B. Array Run
C. Array Formatting
D. Elements

Q10.   The array size exceeds 1000 elements.   Which of the following arrays accommodates this?

A. Multiple arrays
B. Single Multiple arrays
C. Redim Arrays
D. Array Functions

Q11.   What role does Redim Preserve play in arrays?

A. Changes the size of the array
B. Multiples the next array
C. Subtracts the array function
D. None of the above
E. All of the above

Q12.   Arrays can be shared

A. True
B. False

Q13.   Elements of an array can be changed

A. True
B. False

Q14. **Values can be assigned to an array.**

    A. True
    B. False

Q15. **What does 'Redim' in an array stand for?**

    A. The size of which the array is incremented by
    B. The volume of the array
    C. The width of the array
    D. The Length of the array

Q16. **Which of the following retrieves the largest substring of an array?**

    A. Ubound (x)
    B. Redim x[n]
    C. Redim Preserve x[n]
    D. Preserve

Q17. **What does preserve in an array do?**

    A. Elements in the array are maintained
    B. Elements in the array are changed
    C. Elements in the array are not duplicated
    D. Elements in the array are duplicated

Q18. **You have created the following formula. You have also placed the script below into the Select Expert - Record. What will the code enact?**

Formula

```
BooleanVar ResultArray;
StringVar array  arrayExamCode;
if {?Exam Code} = "ALL" then
arrayExamCode := {Exams.ExamCode}
else
arrayExamCode := {?Exam Code};
ResultArray := true;
```

Script

```
StringVar array  arrayExamCode;
```

{Exams.Exam Code} = arrayExamCode

    A. Separates Parameter values
    B. Sets all parameters to false
    C. Sets all parameters to 'Y'
    D. If Parameter prompt is set to All, return all records else return value entered

Q19. An additional certification exam code (RDCR08500a) has been assigned to the current exam certification path. You want to replace the old codes with your new code. Which of the following methods will allow you to add the code and display it on your report?

    A. Local StringVar array ExamCodes := ["RDCR08200a", "RDCR08300a"];
        Redim ExamCodes [3];
        ExamCodes [3] := "RDCR08500a";
        ExamCodes [3]

    B. Local StringVar array ExamCodes := ["RDCR08200a", "RDCR08300a"];
        Preserve Redim ExamCodes [3];
        ExamCodes [3] := "RDCR08500a";
        ExamCodes [3]

    C. Local StringVar array ExamCodes := ["RDCR08200a", "RDCR08300a"];
        Redim Preserve ExamCodes [3];
        ExamCodes [3] := "RDCR08500a";
        ExamCodes [3]

    D. Local StringVar array ExamCodes := ["RDCR08200a", "RDCR08300a"];
        Redim ExamCodes (3);
        ExamCodes (3) := "RDCR08500a";
        ExamCodes (3)

Q20. Which of the following are Array Operators? *(Multiple Answers)*

    A. Make Array
    B. Subscript

C. In Array
D. Redim
E. Preserve
F. Preserve Rem
G. Redim Preserve

Q21.   What is the element limit of array in Crystal Reports?

A. 10
B. 1000
C. 2000
D. 0
E. 1

Q22.   Exams taken by candidates are placed in the same field, i.e. RDCR08200, RDCR08300, RDCR08400. You want the individual exam codes to appear as separate elements. Which of the following will you use?

A. Share()
B. Split()
C. Separate()
D. Divide()

Q23.   Student Exam Records are stored as illustrated in the table below, you want to create a formula which will highlight all students who have completed the required three exams with a status of 'Certified' and 'Not Certified' for all others. Which one of the following formulas will produce the results required?

Results Table

Record No	Exam 1 Of 3	Exam 2 Of 3	Exam 3 Of 3
1	Completed	Completed	Completed
2	Completed		
3			
4	Completed		Completed
5	Completed	Completed	Completed
6	Completed	Completed	Completed
7	Completed	Completed	Completed

A. Local StringVar Array BOCPArray; BOCPArray := [({Results Table.Exam 1 Of 3}), (({{Results Table.Exam 2 Of 3}),(({{Results Table.Exam 3 Of 3})]; UBound (BOCPArray); if UBound (BOCPArray) = 3 then 'Certified' else 'Not Certified'

B. Local StringVar Array BOCPArray; MAKEArray := [({Results Table.Exam 1 Of 3}), (({{Results Table.Exam 2 Of 3}),(({{Results Table.Exam 3 Of 3})]; UBound (BOCPArray); if UBound (BOCPArray) = 3 then 'Certified' else 'Not Certified'

C. Local StringVar PreserveArray BOCPArray; BOCPArray := [({Results Table.Exam 1 Of 3}), (({{Results Table.Exam 2 Of 3}),(({{Results Table.Exam 3 Of 3})]; UBound (BOCPArray); if UBound (BOCPArray) = 3 then 'Certified' else 'Not Certified'

D. Local StringVar Array BOCPArray; BOCPArray := [({Results Table.Exam 1 Of 3}), (({{Results Table.Exam 2 Of 3}),(({{Results Table.Exam 3 Of 3})]; UnBound (BOCPArray); if UnBound (BOCPArray) = 3 then 'Certified' else 'Not Certified'

Q24. **Which of the following formulas will outline 'Early Deliveries or Late deliveries given the date required and the date delivered?**

A. Global DateVar Array DeliveryStatusArray:= [Date({Orders. Required Date}), Date({Orders.Delivered})];
if DeliveryStatusArray[2] < DeliveryStatusArray[1]
then 'Early Delivery'
else 'Late Delivery'

B. B.   Global DateVar Array DeliveryStatusArray:= [Date({Orders.Required Date}), Date({Orders.Delivered})];
if DeliveryStatusArray(2)< DeliveryStatusArray(1)
then 'Early Delivery'
else 'Late Delivery'

C. C.   Global DateVar Array DeliveryStatusArray:= (Date({Orders.Required Date}), Date({Orders.Delivered}));
if DeliveryStatusArray[2] < DeliveryStatusArray[1] then 'Early Delivery' else 'Late Delivery'

CHAPTER 4 - MANAGE REPORTS

Chapter 4 covers questions on exporting reports and the various output file formats, managing projects within the workbench and managing reports within a repository.

Keywords:
Export, Workbench, Formula, Repository, Explorer, PDF, XML, Word, HTML ,Excel, Disconnect

Export a Report

Q1. Which of the following formats are available when exporting a report? *(Multiple Answers)*

A. Acrobat Reader, Microsoft Excel
B. Crystal Reports, Text, ODBC
C. HTML, Report Definition, XML
D. Report Style, Text

Q2. Which one of the following is not an Export Destination available in Crystal Reports?

A. Application
B. Disk File
C. Lotus Domino Mail
D. Exchange Folder
E. Lotus Domino
F. MAPI
G. Folder

Q3. You are exporting a report to Microsoft Excel Data Only format. Which of the following are Microsoft Excel format export options? *(Multiple Answers)*

A. Set Column Width
B. Constant Column width
C. Export object formatting
D. Export Images
E. Main Column alignments

Q4. What do the Use Worksheet Functions for Summaries option indicate?

A. All calculations within the crystal report will be reflected in the exported Excel file
B. All calculations within the report will NOT be reflected in Excel format
C. All objects within the report can be formatted within Excel
D. None of the above

Q5.   You are exporting your report to Microsoft Excel format and you check the Export Object Formatting checkbox. What will happen?

A. All calculations within the report will be reflected in Excel
B. All calculations within the report will NOT be reflected in Excel
C. All objects within the report can be formatted within Excel
D. None of the above

Q6.   You have selected Microsoft Excel Data Only as the export format. Which of the following three options will appear? *(Multiple Answers)*

A. Maximum: Data is exported with no formatting applied
B. Typical: Data is exported with default options applied
C. Minimal: Data is exported with no formatting applied
D. Custom: Data is exported according to selected options

Q7.   You select File – Export. Which of the following options will be available?

A. Export Report
B. Export XML
C. Report Export Options
D. Manage XML Exporting Formats

Q8.   Which of the following export formats will include the reports complete design and formatting details?

A. Definition
B. Report Formula

C. Design

D. Report Definition

Q9. **Which of one the following export application gives the user the option to export the report using the Preview Panel Group as a bookmark?**

A. PDF

B. Microsoft Excel

C. Crystal Reports,

D. Text, ODBC

E. HTML,

F. Report Definition,

G. XML

H. Report Style,

I. Text

Q10. **You want to export the results from your report into a table which other users can connect to. Which of the following apply?**

A. Click the export button and select Microsoft Word – Editable (RTF) as your format and Application as your destination and click ok

B. Click the export button and select Record Style – Columns with space as your format and Application as your destination and click ok

C. Click the export button and select ODBC as your format, select the ODBC format option, (E.g. Microsoft Access, Microsoft Excel) and click ok, select the database to export the reporting data to and click OK, assign a table name and click OK, data can be accessed via the database selected

D. Click the export button and select OLEDB as your format, select the OLEDB format option, (E.g. Microsoft Access, Microsoft Excel) and click ok, select the database to export the reporting data to and click OK, assign a table name and click OK, data can be accessed via the database selected

Manage Reports Using the Workbench

Q1. **What is the main role of the workbench?** *(Multiple Answers)*

A. Organise Reports
B. Delete Reports
C. Edit Reports
D. Access Report

Q2. **Which of the following can be performed using the Workbench?**

A. Publish reports to the Repository
B. Publish reports to Enterprise
C. Publish reports to the Internet
D. Publish reports to the Desktop

Q3. **Which of the following can be viewed within the Workbench?** *(Multiple Answers)*

A. Reports
B. Objects
C. Projects
D. Files

Q4. **Which of the following can be added to the Workbench?** *(Multiple Answers)*

A. Existing Reports
B. Current Reports
C. New Project
D. Object Package

Q5. **Which of the following can be performed within the Workbench?**

A. Check Formulas
B. Check Dependencies
C. Check Objects
D. Check Report

Q6.   A report can be opened within the Workbench?

    A. True
    B. False

Q7.   How can you access the Workbench?

    A. Report - Workbench
    B. File – Workbench
    C. Database – Workbench
    D. View – Workbench

Q8.   You want to add a new project to the Workbench. Which of the following apply?

    A. View – Workbench – Add – Add New Project
    B. File – Workbench – Add – Add New Project
    C. Database– Workbench – Add – Add New Project
    D. Report – Workbench – Add – Add New Project

Q9.   You want to add a new Object Package to the Workbench. Which of the following apply?

    A. Right-click the project folder and select – Add New Object Package – Enterprise Connectivity is required
    B. Right-click the project folder and select – Add New Object Package – Enterprise Connectivity is NOT required

Q10.   By adding a new project to the Workbench a New Folder is created.

    A. True
    B. False

Q11.   You want to save your report within the Workbench to Business Objects Enterprise, which method applies?

    A. Within the Workbench right-click the report object and select Save to Business Objects Enterprise, log into Enterprise and save to required folder
    B. Within the Workbench right-click the report object and select

Add Current Report, log into Enterprise and save to required folder

C. Reports cannot be saved from the workbench to Business Objects Enterprise, log into Enterprise and save to required folder

Manage reports using the repository

Q1.   A Repository is defined as which of the following?

A. The administrative console for report filing
B. Crystal Management Server
C. Central database for storing and sharing codes, Custom Functions, Text Objects, Logos, SQL Commands, and Business Views
D. Central Management Server

Q2.   You want to view the contents of the Repository which of the following apply?

A. Field Explorer
B. Custom Function Explorer
C. Report Explorer
D. Enterprise Explorer
E. Repository Explorer
F. Repository Expert Database

Q3.   Which of the following methods allows you to access the Repository Explorer within a report? *(Multiple Answers)*

A. Choose view | Field Explorer | Repository Explorer
B. Choose view | Repository Explorer
C. Section Expert | Repository Explorer
D. Select Expert| Repository Explorer
E. Database Explorer | Repository Explorer
F. Click the Repository icon

G. Click the Repository icon

Q4.   Which of the following applies to folder creation within the Repository?

A. Repository Folders are limited to single folders.
B. A maximum of 7 folders apply to the Repository folder.
C. Repository sub-folders cannot be created within the Repository Explorer

D. A set of Folders and Subfolders can be created within the Repository

Q5. **You want to rename a folder that already exists in the Repository Explorer, which of the following three describes the best procedure to follow?** *(Multiple Answers)*

A. Select the folder, hold the mouse button down on the folder for a few seconds and rename when the folder becomes editable
B. Select folder and click F2 and rename
C. Right-click the folder and choose rename from the pop-up menu
D. Select the folder from the Report Expert and rename

Q6. **Which of the following are Repository Explorer position options?** *(Multiple Answers)*

A. Undock
B. Autohide
C. Close
D. Open

Q7 **You place your cursor over the Repository Explorer. What will happen?**

A. The mouse icon will disappear
B. The mouse icon will turn to an hour glass icon
C. The Tool Tips will appear
D. None of the above

Q8. **Which of the following cannot be seen through the Repository Explorer?**

A. Text Objects
B. Bitmap (pictures)
C. Custom Functions
D. SQL Commands/Queries

Q9. **You want to see the creator of all objects within the Repository. Which of the following options should be selected?**

A. Change View settings
B. Change Filter settings
C. Advanced Filtering
D. Change Repository settings

Q10. **Which of the following applies to Repositories?** *(Multiple Answers)*

A. Changes and updates can be made and saved to the Repository for shared use
B. The Repository can only be accessed by logging on to Business Objects Enterprise
C. User authentication is not required
D. Code sharing is one of the advantages of the Repository
E. Reduces design time due to its sharing capability

Q11. **You do not have Crystal Enterprise installed, you try to access the repository explorer; this action will not be permitted.**

A. True
B. False

Q12. **How can you determine which objects within the Repository are shown and how they are arranged?**

A. Change Repository Settings
B. Change File Settings
C. Change View Settings
D. Change Filter Settings

Q13. **When a report is opened you want to ensure all repository objects are updated. What should you do?** *(Multiple Answers)*

A. Select File from the toolbar – click on Repository options – select the Repository Tab – check the update connected Repository objects when loading reports

B. Select File from the toolbar – click on Report options – select the Repository Tab – check the update connected Repository objects when loading reports

C. Select File from the toolbar – click on Summary info – select the Repository Tab – check the update connected Repository objects when loading reports Select File from the toolbar – click on options – select the Repository Tab – check the update connected Repository objects when loading reports

D. Select File from the Menu Bar – click on options – select the reporting Tab –under Enterprise Settings check Update Connected Repository Objects on Open

E. Select File – Save As, select the Enterprise Folder, within Enterprise select the required folder and tick the Enable Repository Refresh checkbox

Q14. To view subfolders within the repository which of the following should you select?

A. Expand All
B. Refresh Folders
C. Open All
D. Refresh All

Q15. Deleted Repository Objects will still remain within the report.

A. True
B. False

Q16. How can you log off the Repository?

A. Right-click the Repository and select Log Off Server
B. Right-click the top folder within the Repository and select Log Off Server
C. Press the F4 key
D. Press the F12 key

Q17. You want to delete objects in the Repository. Which of the following applies? *(Multiple Answers)*

A. You must delete all objects and subfolders within the folder, before you delete the folder
B. Right-click the empty folder and select delete
C. Right-click the folder, select save as, this will save the folder and it's contents to the C:\ drive then right click again and delete

Q18. You have selected the option to view the Repository Explorer, but you are not logged on. Which of the following should be applied to enable this process?

A. Press the F5 and the Repository will automatically be activated
B. Click the logon icon and enter your authentication details
C. Click the Repository icon, this should log the user on automatically
D. Select database Repository Explorer and log on

Q19. A Repository object must be disconnected before it can be modified?

A. True
B. False

Q20. You want to add a Custom Function to the Repository. Which of the following methods apply? *(Multiple Answers)*

A. Drag and Drop the Custom Function from the Field Explorer into the Repository
B. Right-Click the Custom Function within the Formula Workshop and 'Add to Repository'
C. Drag and Drop the Custom Function from the Report Explorer into the Repository
D. Drag and Drop the Custom Function from the File Explorer into the Repository
E. Drag the Custom Function from the Report Custom Functions node within the Formula Workshop Tree and drop into a Repository Custom Functions node.

Q21. **You want to amend the company logo which is saved to the Repository. Which of the following actions must you do?**

A. Right-click the object on the report in the Formula Explorer, choose disconnect from the Repository, amend object, then right-click object again and add to Repository with the same name.
B. Right-click the object on the report in the Repository Explorer, choose disconnect from the Repository, amend object, then right-click object again and add to Repository with the same name.
C. Right-click the object on the report in the Database Explorer, choose disconnect from the Repository, amend object, then right-click object again and add to Repository with the same name.
D. Right-click the object on the report in the Report Explorer, choose disconnect from the Repository, amend object, then right-click object again and add to Repository with the same name.
E. Right-click the object, choose disconnect from the Repository, amend object, then right-click object again and add to Repository with the same name.

Q22. **You are unable to modify an object saved to the repository. What is the likely cause of this problem?**

A. The object cannot be modified as it has been created with PaintshopPro and must be resized within the PaintShopPro environment
B. The object cannot be modified as it is still connected to the shared Repository and therefore needs to be disconnected before the amendment can be made.
C. Right-click the object and tick modification allowed

Q23. **Folders within the Repository cannot be renamed.**

A. True
B. False

Q24. **Which of the following apply to the Repository?** *(Multiple Answers)*

A. Drag and drop object images into the Repository

B. Custom Functions can be added via the Formula Workshop
C. SQL Commands can be added through the Database Expert.
D. Reports can be dragged into the Repository

Q25. **You want to add a Text Object within your report to the Repository. Which of the following apply?**

A. Right-click the Text Object and click edit Repository
B. Right-click the Text Object and select add to Repository from the drop-down menu, enter the authentication details and enter the required descriptions and save to the appropriate folder
C. Right-click the Text Object and click save to Repository
D. Right-click the Text Object and click Format Graphic and Save to Repository

Q26. **You place the mouse over a SQL Command in the Repository. Which of the following will appear?**

A. Repository Name
B. Script creation date
C. File Name
D. Description and Script

Q27. **You want to add a SQL Command to the Repository. Which of the following apply?**

A. SQL Commands can be added through the Explorer Expert, by clicking the 'Add to Repository' Checkbox from the ADD COMMAND section
B. SQL Commands can be added through the Report Expert, by clicking the 'Add to Repository' Checkbox from the ADD COMMAND section
C. SQL Commands can be added through the Database Expert, by selecting 'Add Command with the database connection and entering the SQL script within the Add Command to Report dialog section, clicking ok and selecting Add to Repository'
D. SQL Commands can be added through the Section Expert, by clicking the 'Add to Repository' Checkbox from the ADD COMMAND section

Q28. Which of the following scripts will produce an error message when entered within the Add Command?

A. SELECT Reports.* FROM Reports;
B. SELECT Reports FROM Reports
C. select * from Reports
D. None of the above

Q29. Given the query below in the SQL Command, you want to add a parameter to replace the Candidate_No. Which of the following apply?

Select [Examhints].Candidate_No, [Examhints].Candidate_FName,
[Examhints].Candidate_LName,   [Examhints].Candidate_Address,
[Examhints].Candidate_Score
From [Examhints]

A. Click the parameter button and enter the parameter details as follows: parameter name: Candidate_No, prompting text: Enter Candidate_No, value type: String and default value: RDCR08201, highlight the exam_id within the parameter list and replace the exam_id with the parameter created, users will be prompted for an Candidate_No when using the SQL Command created.
B. Click the modify button and enter the parameter details as follows: parameter name: Candidate_No, prompting text: Enter Candidate_No, value type: String and default value: RDCR08201, highlight the exam_id within the parameter list and replace the exam_id with the parameter created, users will be prompted for an Candidate_No when using the SQL Command created.
C. Click the Command button and enter the parameter details as follows: parameter name: Candidate_No, prompting text: Enter Candidate_No, value type: String and default value: RDCR08201, highlight the exam_id within the parameter list and replace the exam_id with the parameter created, users will be prompted for an Candidate_No when using the SQL Command created.
D. Click the create button within the SQL Command dialog box and enter the parameter details as follows: parameter name: Candidate_No, prompting text: Enter Candidate_No, value type: String and default value: RDCR08201, highlight the exam_id within the parameter list and replace the exam_id with the parameter created, users will be prompted for an Candidate_No when using the SQL Command created

CHAPTER 5 -CREATE AN ADVANCED REPORT

Chapter 5 covers questions on creating advanced reports; this includes parameter creation, application and formatting, using dynamic cascading, building cross-tabs, running totals, report alerts, Top N and Bottom N reports.

Keywords:
Static, Dynamic, Edit Mask, Top N, Bottom N, Cross-tabs, Cascading, Running Totals, Alerts.

Create a Parameter

Q1: Which of the above named ID's in fig 5.0 will appear when "AAAA" is placed in the Edit Mask? *(Multiple Answers)*

A. 8
B. 12
C. 16
D. 4

Q2: Which one of the ID's in fig 5.0 will appear when "aaaaaa" is placed in the Edit Mask? *(Multiple Answers)*

A. 7,8,9,10,11,12,13,14,15,16,22,25
B. 1,2,3,4,5,6
C. 19,20
D. 4

Q3: Which one of the report ID's in fig 7.0 will appear when "######" is placed in the Edit Mask?

A. 7
B. 24
C. 8
D. 9

Sample Data - *Fig.5.0*

ID	Value
1	RDCR08201
2	RDCR08301
3	RDCR08401
4	RDCR201
5	RDCR301
6	RDCR401
7	RE
8	7789
9	1
10	20
11	3
12	CASE
13	REPEAT
14	EVALU1
15	08
16	2003
17	BOCPRT1
18	BOCPRPT3
19	Report Version
20	Antonia Iroko
21	20092008
22	RPT
23	778883
24	HINTS 3
25	comply

Q4: Which of the following report ID's in fig 7.0 will appear when "????" is placed in the Edit Mask?

A. 9
B. 12
C. 11
D. 10

Q5: Which one of the report ID's in fig 7.0 will appear when "0" is placed in the Edit Mask? *(Multiple Answers)*

A. 24
B. 22
C. 9
D. 11

Q6: Which of the report ID's in fig 7.0 will appear when "&&&&&&&" is placed in the Edit Mask? *(Multiple Answers)*

A. 4,5
B. 5,3
C. 6,17
D. 24

Q7: The Edit Mask can be applied to which of the following data types?

A. Date
B. Currency
C. DateTime
D. String
E. Boolean
F. Time
G. Number

Q8: Which of the following has been placed in the Edit Mask to produce the following?

Enter a Value:

●●●●●●●●●●●●●●

A. >
B. <
C. *******
D. Password
E. Security
F. ;

Q9: **Which of the following signs precedes the parameter?**

A. $
B. //
C. :
D. ?
E. @
F. %

Q10: **Which of the following can be used to populate the parameter list with default values?** *(Multiple Answers)*

A. Append all database value
B. Enter default values manually
C. Import a pick list
D. Populated from another report

Q11: **You want to be prompted with the parameter description. Which of the following settings apply?**

A. Prompt With Description = True
B. Prompt With Description Only = True
C. Prompt With Description = False
D. Prompt With Description Only = False

Q12: **Which of the following options need to be set to allow users to enter customized values?**

A. Set Allow Multiple Values To True
B. Set Discrete Values To True
C. Set Customize Values To True
D. Set Allow Custom Values To True

Q13: **Which of the following options need to be set to allow users to enter a start and end date?**

A. Uncheck the Allow Multiple Values in the edit parameter field
B. Click the range values radio button
C. Check the Allow Multiple values
D. Set Allow Multiple Values to true
E. Click the discrete and range values radio button

Q14: **Users of your report are required to enter a begin date, and end date. When prompted. Which of following parameter options apply?**

A. Allow discrete values
B. Allow multiple values
C. Allow range values
D. Allow custom values
E. Min Length
F. Max Length

Q15: **Which of the following are mandatory to ensure that a parameter works?** *(Multiple Answers)*

A. Parameter Name
B. Static or Dynamic Parameter
C. Prompting Text
D. Value Type
E. Allow Multiple\Discrete\Range Values
F. Default Value
G. Description of default values

Q16: **Which parameter value can be displayed on a report without further formula coding?**
A. Range
B. Discrete
C. Multiple values
D. Custom Values

Q17: **You are prompted to Use Current Parameter Values, or Prompt For New Parameter Values. You choose the latter. Which of the following apply?**

A. The report will refresh immediately form the database
B. A new set of parameters values will be required after which the report will refresh directly from the database
C. The report will produce a database error message
D. The report will use the previous parameter values entered to refresh the report

Q18: **You want to prevent users from entering sales values less than 2000, or greater than 100000, when running the sales report. Which of the following apply?**

A. In the options settings section of the edit parameter dialog box set the Min value to 2001 and the Max value to 100000
B. In the options settings section of the edit parameter dialog box set the Min value to 2001 and the Max value to 99999
C. In the options settings section of the edit parameter dialog box set the Min Value to 2000 and the Max Value to 100000
D. In the options settings section of the edit parameter dialog box set the Min value to 2002 and the Max value to 99999

Q19: **How are parameters integrated in reports?** *(Multiple Answers)*

A. Select Expert
B. Group Selection
C. Record Selections
D. Report Busting indexes

Q20:   **Parameters can be used for grouping.**

    A. True
    B. False

Q21:   **Parameter fields can be dragged and dropped from the Field Explorer, onto the report.**

    A. False
    B. True

Q22.   **You want the give the users the option not to enter any value in the customer name prompt when running a report. Which of the following apply?**

    A. This is not possible, a value must be entered in all parameter prompts
    B. Set the Optional Editor to False
    C. Set the Edit mask to ***
    D. Set the Optional Prompt to True

Q23:   **You have created a parameter named Report_Numbers. You want to query the report table to retrieve report numbers between 1 and 115. The parameter type is a Number. From the Select Expert you select the field report_ number equals, and look for the parameter in the drop down list. The parameter does not appear in the drop down list. Which of the following reasons apply?**

    A. The data types of the parameter and the database field are different
    B. You have not set Allow Multiple Values to true
    C. You have not set allow range vales to true
    D. You have not set allow discrete vales to true

Q24:   **Which of the following will allow you to integrate a range parameter within the Select Expert?**

    A. {report.number} in {?Report_Number}
    B. {report.number} = {?Report_Number}
    C. {report.number} in {?Report_Number}to {?Report_Number}
    D. not ({report.number} in [{?Report_Number}])

Q25: **You have created a Range Parameter, which you have applied to the archive number. When refreshing the report, you enter a range between 300 and 400. You uncheck the 'include this value' checkbox for the start range, which you have set as 300, and uncheck the include value checkbox for the 'end range', which you have set as 400. Which of the following is true?**

A. Values over 200 will be included (plus 200 itself), and values up to 400 will be included
B. Values over and including 200 will be included and values up 400 but not 400 will be included
C. Values over 200 will be included (but not 200 itself), and values up to 400 will be included (but not 400 itself)
D. Values over 200 will be included (but not 200 itself), and values up to 400 will be included

Q26: **You have specified a range for your parameter, with a start value of 10000, and an end value of 12999. You tick the no lower value checkbox for the start range. Which of the following values will appear?** *(Multiple Answers)*

A	10
B	12888
C	10601
D	14909
E	10000

Q27: **You have created a static parameter however recently updated database values do not appear in the dropdown list. Which of the following reasons apply?**

A. Static Parameters must be updated by selecting Action – and Append all database values
B. The report must be refreshed to pick up the list
C. The report was saved with data and therefore needs to be refreshed to include new figures in the pick list
D. The pick list will be updated after several refreshes

Q28: Which of the following can be used to display a Multiple Parameter?

A. Financial Function
B. Control Structure
C. Array functions
D. Range operators

Q29: You have created a parameter and a formula with the same name. The formula will fail to run due to a naming conflict

A. True
B. False

Q30: You want to create a parameter which will only show a list of exams applicable to a selected certification type. What is the name of this parameter type?

A. Dynamic and Cascading
B. Static and Cascading
C. Edit Mask and Cascading

Q31: You refresh your report using the Use Current Parameter Values. Which of the following apply?

A. The report will refresh immediately from the database
B. You will be prompted for a new set of parameter values after which the report will refresh directly from the database
C. The report will produce a database error message
D. The report will use the previous parameter values entered to refresh the report

Q32: Which of the following options are available under 'Actions' within the parameter' dialog box? *(Multiple Answers)*

A. Append all database values
B. Export
C. Import
D. Clear

Q33: **You want to set the order in which parameters appear in Business Objects Enterprise. Which of the following apply?**

A. Right-click the parameter fields and select Order and change the parameter order using the arrows
B. Highlight one parameter field and use the arrow to move the parameter to the required position
C. Right-click the parameter fields and select Set Parameter Order and change the parameter order using the arrows

Q34: **You have created two Static Parameters based on then Exam_ID and Candidate Name, you want to be prompted for the Exam_ID first and the Candidate Name second, how would you implement this format?**

A. Highlight the parameter and press the arrow on your keyboard
B. Right-click the parameter and select set parameter order, use the arrow to achieve the order required
C. Integrate parameter within the Select Expert in the order required
D. This cannot be done once the parameters have been created

Q35: **What is the difference between a Static and Dynamic Parameter?** *(Multiple Answers)*

A. A Static Parameter's picklist is not directly linked to the database and must be updated manually
B. A Dynamic Parameter's picklist is directly linked to the database field and is up to date with current database values
C. A Dynamic Parameter can be designed as a cascading prompt, giving the user the option to select only the values applicable to the initial value selected
D. A Static Parameter are linked directly to the database field

Q36. **You have created and integrated a Static Parameter based on the exam_id field, when the report is refreshed there are no values present in the picklist. Which of the following can be used to populate a static parameter list?**

A. Append all database values
B. Attach database fields
C. Insert database fields
D. Add database fields

Q37. **Which of the following formulas will give the user the option to select all parameter values from a picklist when prompted or a specific value?**

In this case the field name is {Customer.City} and the parameter is {?SelectClientCity}

A. In the Select Expert enter : {?SelectCustomerCity} ='All' or {Customer.City} = {?SelectCustomerCity})
B. In the Select Expert enter : {?SelectCustomerCity} =" or {Customer.City} {?SelectCustomerCity})
C. In the Select Expert enter : {?SelectCustomerCity} ={Customer.City} or {Customer.City} ={?SelectCustomerCity})

Q38. **You want to create a manual date parameter using the subscription date which will allow report users to enter dates manually without using the calendar. Which of the following formulas will work?**

A. Create a Starts and End Date parameter using a string as the data type, enter a default value of 'All' in both parameters. Create the following formula using the Subscription Date:
(@SubDate}
mid(totext({SubscriptionDate }),7,4) + ""
+mid(totext({SubscriptionDate }),4,2) + "" +
mid(totext({SubscriptionDate }),1,2)   Enter the following formula in the Select Expert
{?StartDate} to {?EndDate} = 'All' or (@SubDate}in
   mid({?StartDate},7,4) + "" + mid({?StartDate},4,2) + ""
   + mid({?StartDate},1,2) to mid({?EndDate},7,4) + "" +
   mid({?EndDate},4,2) + "" + mid({?EndDate},1,2)
B. Create a Starts and End Date parameter using a string as the data type, enter a default value of 'All' in both parameters. Create the following formula using the Subscription Date: (@SubDate}
Cdate(SubscriptionDate) Enter the following formula in the Select Expert
{?StartDate} to {?EndDate} = 'All' or (@SubDate}in
   mid({?StartDate},7,4) + "" + mid({?StartDate},4,2) + ""
   + mid({?StartDate},1,2) to mid({?EndDate},7,4) + "" +
   mid({?EndDate},4,2) + "" + mid({?EndDate},1,2)

Q39. You want to give the users the option not to enter any value in the customer name prompt when running a report. Which of the following apply?

A. This is not possible, a value must be entered in all parameter prompts
B. Set the Optional Editor to False
C. Set the Edit mask to ***
D. Set the Optional Prompt to True

Q40. You have set the optional prompt to true, how will the field associated with this setting be represented in the Select Expert – Record ?

A. (Not HasValue({Table.fieldname}) OR {Table.fieldname} ={Parameter})
B. (HasValue({Table.fieldname}) OR {Table.fieldname} ={Parameter})
C. (Not Value({Table.fieldname}) OR {Table.fieldname} ={Parameter})

Q41. Parameter name fields are limited to 255 alphanumeric characters only

A. True
B. False

Q42. You have created a pick list (LOV) based on the customer's city, several regions appear within the dropdown list hence the pick list is split into several batches, you are only interested in viewing a LOV with 'Ch'. Which method will achieve this?

A. When prompted to enter values, click the set Filter option and enter Ch and click ok
B. Filtering is not available within the Parameter Prompt

Q43. Parameters (LOV) are split into batch sizes depending on the number of values in the list?

A. False
B. True

Q44.   What does LOV stand for?

A. Line of values
B. List of Values
C. Loop of Values

Use Dynamic Cascading Prompting

Q1. Which of the following best describes a parameter cascading list of values?

    A. Narrows a parameter pick list down to an applicable selection once an initial selection is made.

    B. Provides a static list of values which are applicable to the initial selection made

Q2. A cascading list of values can only be created in conjunction with Dynamic Parameters?

    A. True

    B. False

Q3. You have received a request to create a report which will allow the user to select a certification route of interest and view the exams required and the pass score. How would you create the parameter?

    A. Right- click parameter fields and select New –enter the parameter name – enter prompt group text and select Static from the list values, click add an item and select certification from the database table, click to create parameter, repeat the same procedure for the exam id and score. The Dynamic parameter will be Certification\ Examid \Score

    B. Right- click parameter fields and select New –enter the parameter name – enter prompt group text and select Dynamic from the list values, click add an item and select certification from the database table, click to create parameter, repeat the same procedure for the exam id and score. The Dynamic parameter will be Certification\ Examid \Score

Q4. You have created a parameter with a cascading list of values, using the certification as the leading prompt, followed by the exam id and score. Which of the following statements is true?

    A. When prompted the user must select values from the certification parameter before other prompts can be activated or selection

B. When prompted the user can select any of the parameter values first

Q5.   You have created a cascading parameter for Country and City, hence when a user selects a Country all the cities within that country will appear in the city prompt, you have set the country parameter as an optional prompt and not the city. Which of the following statements is true?

A. The optional prompt will apply to both parameters as they are cascading parameters
B. Optional prompts cannot be set for cascading parameters

Build and Format a Basic Cross-Tab

Q1. **Which of the following defines a Cross-Tab?** *(Multiple Answers)*

A. Cross-Tabs consist of rows and columns and are used to present data in a logical format

B. Cross-Tabs can only be created when groups exist within the report

C. Cross-Tabs are Excel based sheets imported into a report

D. Fields within the Cross-Tab can be summarized based on sums, counts, distinct counts and all other summaries applicable to a field type

Q2. **You want to ensure objects in the Cross-Tab are not being printed over each other. What should you do?**

A. Select Field Expert, highlight the Group Header where the Cross-Tab is placed and place a tick in the check box or Relative Position

B. Select Selection Expert, highlight the Group Header where the Cross-Tab is placed and place a tick in the check box or Relative Position

C. Select Group Expert, select the Common Tab and highlight the Group Header where the Cross-Tab is placed and place a tick in the check box or Relative Position

D. From the Section Expert, select the Common Tab and highlight the Group Header where the Cross-Tab is placed and place a tick in the Relative Position check box

Q3. **You want to remove all Calculated Members from your Cross-Tab. Which of the following apply?** *(Multiple Answers)*

A. Right-Click the Cross-Tab and select Advanced Calculations – Calculated Member, Highlight the calculated member within the Cross-Tab Calculated Member Expert and select Remove

B. Right-Click the Calculated Member within the Cross-Tab and select delete

C. Right-Click the Cross-Tab and select Calculated Member, Highlight the calculated members and select Remove

D. Right-Click the Calculated Member and select Advanced Calculations and select Remove

Q4.   You want to insert a chart based on the values of your Cross-Tab. Which of the following apply?

A. Right-click the Cross-Tab and select Insert Chart
B. From the menu – select insert Chart
C. Right-click the Cross-Tab and select Convert Cross-Tab to Chart
D. From the menu – select Convert Cross-Tab to Chart

Q5.   You want to reduce the space within your Cross-Tab as illustrated below (Diagram A), how can you change it to appear as illustrated in (Diagram B)?

A. Right-click the Cross-Tab and select Show Margins
B. Right-click the Cross-Tab and select Format Grid Lines
C. Right-click the Cross-Tab and select Grid Options and un-tick Show Cell Margins
D. Right-click the Cross-Tab and select Grid Options and tick Show Cell Margins

	Warsaw	West Yorkshire	WI	WV
Xtreme Adult Helmet	£11,310.45	£0.00	£3,834.05	£0.00
Xtreme Anatomic Ladies Saddle	£3,409.77	£0.00	£6,149.64	£0.00
Xtreme Anatomic Mens Saddle	£0.00	£0.00	£5,938.40	£0.00
Xtreme Gellite Ladies Saddle	£95.51	£23.50	£3,526.59	£23.50
Xtreme Rhino Lock	£0.00	£0.00	£9,169.36	£0.00
Xtreme Wide MTB Saddle	£0.00	£0.00	£43.50	£0.00
Xtreme Youth Helmet	£83.40	£0.00	£1,938.57	£0.00

Diagram A

	Warsaw	West Yorkshire	WI	WV
Xtreme Adult Helmet	£11,310.45	£0.00	£3,834.05	£0.00
Xtreme Anatomic Ladies Saddle	£3,409.77	£0.00	£6,149.64	£0.00
Xtreme Anatomic Mens Saddle	£0.00	£0.00	£5,938.40	£0.00
Xtreme Gellite Ladies Saddle	£95.51	£23.50	£3,526.59	£23.50
Xtreme Rhino Lock	£0.00	£0.00	£9,169.36	£0.00
Xtreme Wide MTB Saddle	£0.00	£0.00	£43.50	£0.00
Xtreme Youth Helmet	£83.40	£0.00	£1,938.57	£0.00

Diagram B

Q6. You Right-click the Cross-Tab illustrated in fig 7.2 below and select Pivot Cross-Tab. Which of the following will happen?

	London	Birmingham	Manchester
Crystal Reports 2008	200000	40000	100000
Crystal Reports XI	7000	10000	430
Crystal Reports XI R2	400988	1000	722
Crystal Reports 10	20000	577	211
Crystal Reports 8.5	7000	9000	0
Crystal Reports 7	12000	100	76000

Fig: 7.2

   A. Results will be displayed in a Vertical Format, Product Names will appear as a column and regions will appear in the row
   B. Results will be displayed upside down
   C. Columns will be displayed as Rows and Rows as Columns
   D. None of the above

Q7. How can you display the name of the summarized field within your Cross-Tab?

   A. Right-click the Cross-Tab and select Cross-Tab Expert and select Grid Options and Show Summarized Field Labels
   B. Right-click the Cross-Tab field and select Summarized Field Labels - Show Summarized Field Labels
   C. Right-click the Cross-Tab and select Format Cross-Tab and select Summarized Field Labels and Show Summarized Field Labels
   D. Right-click the Cross-Tab and select Cross-Tab Expert and select Group Sort Expert and Show Summarized Field Labels

Q8. Totals appear on the top left section of the Cross-Tab. You want to change the location to the bottom of the Cross-Tab. Which of the following apply?

   A. Right-click the Total text within the Cross-Tab and select Row Grand Totals and select Totals On Top
   B. Right-click the Total text within the Cross-Tab and select Grand Totals and select Totals on Top
   C. Right-click the Total text within the Cross-Tab and select Totals and select Totals on Top
   D. Right-click the Total text within the Cross-Tab and select move Grand Totals to Top

Q9. **Which of the following procedures apply to building a Cross-Tab?**

A. From the menu select View – Cross-Tab, a Cross-Tab object will appear, place the box in the Report Header, Right-click the Cross-Tab box and select Cross-Tab Expert, the Cross-Tab dialog box will appear, from the available fields section – select field required for your columns using the arrow, a field for your Row and a value for your summarized fields and click Ok

B. From the menu select Insert – Cross-Tab, a Cross-Tab object will appear, place the box in the Report Header, Right-click the Cross-Tab box and select Cross-Tab Expert, the Cross-Tab dialog box will appear, from the available fields section – select the field required for your columns using the arrow, and the field for your Row and a value for your summarized fields and click Ok

C. From the menu select Report - Insert – Cross-Tab, a Cross-Tab object will appear, place the box in the Report Header, Right-click the Cross-Tab box and select Cross-Tab Expert, the Cross-Tab dialog box will appear, from the available fields section – select field required for your columns using the arrow, a field for your Row and a value for your summarized fields and click Ok

D. From the menu select Insert – Chart- Chart type- Cross-Tab, a Cross-Tab box will appear, place the box in the Report Header, Right-click the Cross-Tab box and select Cross-Tab Expert, the Cross-Tab dialog box will appear, from the available fields section – select field required for your columns using the arrow, a field for your Row and a value for your summarized fields and click Ok

Q10. **Which of the following apply when creating a new Cross-Tab report?**

A. Select File – New – Cross-Tab Chart – Create a New Data Connection – Select the required tables from the available Data source – Assign fields to the Column. Rows and summarized sections of the Cross-Tab and click next, No chart, apply filter if applicable – choose Cross-Tab style and click finish

B. Select File – New – Cross-Tab Report – Create a New Data Connection using the Cross-Tab Report Creation Wizard – Select the required tables from the available Data source – Assign fields to the Column. Rows and summarized sections of the Cross-Tab and click next, No chart, apply filter if applicable – choose Cross-Tab style and click finish

C. Select File – New – Cross-Tab Dialog – Create a New Data Connection – Select the required tables from the available Data source – Assign fields to the Column. Rows and summarized sections of the Cross-Tab and click next, No chart, apply filter if applicable – choose Cross-Tab style and click finish

D. Select File – New – Cross-Tab Expert Report – the Cross-Tab Report Creation Wizard dialog box will appear - Create a New Data Connection – Select the required tables from the available Data source – Assign fields to the Column. Rows and summarized sections of the Cross-Tab and click next, No chart, apply filter if applicable – choose Cross-Tab style and click finish

Q11. You want to display the TopN sales per city within your Cross-Tab. Which of the following apply? *(Multiple Answers)*

A. Right-click the Cross-Tab and select Group Sort Expert, for Top N based on the quantity where N is 5 and click ok

B. Select Report - Group Sort Expert, select Top N based on sales amount where N is 5 and uncheck include others and click ok

C. Select Report – Cross-Tab, Group Sort Expert, for Top N based on the quantity where N is 5 and uncheck include others and click ok

D. The Group Sort Expert cannot be applied to a Cross-Tab

Q12. You want to display the total discount awarded to each city and the name of the customer within the city.

A. Right-click the Cross-Tab and select Cross-Tab Expert, add the discount field to the summarized section and add the Customer Name to the Row and click ok

B. Right-click the Cross-Tab and select Format Cross-Tab Expert, add the discount field to the summarized section and add the Customer Name to the Row and click ok

C. Right-click the Cross-Tab and select Edit Cross-Tab, add the discount field to the summarized section and add the Customer Name to the Row and click ok

D. Right-click the Cross-Tab and select Format Expert, add the discount field to the summarized section and add the Customer Name to the Row and click ok

Q13. Which of the following sections can you place a Cross-Tab? *(Multiple Answers)*

A. Report Header
B. Group Header
C. Details section
D. Group Footer
E. Report Footer

Q14. Which one of the following statements is true?

A. Charts and Cross-Tab objects placed in the Page Header area print at the beginning of each page
B. Charts or Cross-Tabs cannot be placed in the page header section
C. Charts and Cross-Tabs cannot be placed in the Group Header section
D. Charts or Cross-Tabs can be placed in the page header section
E. Charts and Cross-Tabs can be placed in the detail section

Q15. The Cross-Tab chart is only available when Cross-Tab objects already exist in your report.

A. True
B. False

Q16. Which one of the following is true?

A. A Cross-Tab is not based on an existing group or its summarized fields
B. A Cross-Tab requires a group or summary fields
C. A Cross-Tab requires a group
D. A Cross-Tab requires summary fields with the main report

Q17. You want to highlight the background of all orders less than or equal to 7000 with red and sales of 10000 and over with blue. This will enable the Sales Team to identify customers who will be awarded a discount on their next orders. Which of the following apply?

A. Right-click the summarized filed within the Cross-Tab and select Format Field, select the Border Tab and click the formula button beside the background and enter the following formula: if CurrentFieldValue <= 7000 then Crred else CrBlue

B. Right-click the summarized filed within the Cross-Tab and select Format Field, select the Font tab and click the formula button beside the background and enter the following formula: if CurrentFieldValue <= 7000 then Crred else CrBlue

C. Right-click the summarized filed within the Cross-Tab and select Format Field, select the Number tab and click the formula button beside the background and enter the following formula: if CurrentFieldValue <= 7000 then Crred else CrBlue

D. Right-click the summarized filed within the Cross-Tab and select Format Field, select the Common Tab and click the formula button beside the background and enter the following formula: if CurrentFieldValue <= 7000 then Crred else CrBlue

Q18. **You want to display the text discount awarded for all sales equal to 51 and over and no discount for 50 and less. Which of the following apply?**

A. Right-click the summarized filed within the Cross-Tab and select Format Field, select the Border Tab and click the formula button beside the background and enter the following formula: if CurrentFieldValue <= 50 then No Discount else Discount Awarded

B. Right-click the summarized filed within the Cross-Tab and select Format Field, select the Font tab and click the formula button beside the background and enter the following formula: if CurrentFieldValue <= 50 then No Discount else Discount Awarded

C. Right-click the summarized filed within the Cross-Tab and select Format Field, select the Number tab and click the formula button beside the background and enter the following formula: if CurrentFieldValue <= 50 then No Discount else Discount Awarded

D. Right-click the summarized filed within the Cross-Tab and select Format Field, select the Common Tab and click the formula button beside the Display String and enter the following formula: if CurrentFieldValue <= 50 then 'No Discount' else 'Discount Awarded'

Q19. You want to change the current style of the Cross-Tab. Which of the following methods apply?

A. Right-click the Cross-Tab and select Cross-Tab Expert, click the Style Tab and select the style required
B. Highlight the Cross-Tab and select format from the Menu Bar and select Cross-Tab Expert, click the Style Tab and select the style required
C. Select Cross-Tab from the Menu Bar and select Cross-Tab Expert, click the Style Tab and select the style required
D. Select view Cross-Tab from the Menu Bar and select Cross-Tab Expert, click the Style Tab and select the style required

Q20. You want to apply your own style to a Cross-Tab Expert. What can you do to create your own style?

A. Highlight the Cross-Tab, select format from the Menu Bar and select Cross-Tab Expert, click the Customize Style Tab and create the style required
B. Select Cross-Tab from the Menu Bar and select Cross-Tab Expert, click the Style Tab and select the style required
C. Select view Cross-Tab from the Menu Bar and select Cross-Tab Expert, click the Style Tab and select the style required

Q21. A Cross-Tab can be resized by highlighting the field and dragging it to its required style

A. True
B. False

Q22. There are several empty columns in your Cross-Tab. Which of the following methods will enable you to remove these columns? *(Multiple Answers)*

A. Select Cross-Tab from the Menu Bar and select Cross-Tab Expert, click the Style Tab and select the style required
B. Select View Cross-Tab from the Menu Bar and select Cross-Tab Expert, click the Style Tab and select the style required
C. Select format from the Menu Bar and select Cross-Tab Expert, click the Customize Style Tab and check the Suppress Empty Columns

D. Right-click the Cross-Tab and select Cross-Tab Expert, click the Customize Style Tab and check the Suppress Empty Columns

Q23. Formulas can be added to Cross-Tabs

A. True
B. False

Q24. You want to remove all cell margins within the Cross-Tab. Which of the following apply?

A. Right-click the Cross-Tab and select Grid options and click remove cell margins, the tick will disappear
B. Right-click the Cross-Tab and select Grid options and click Show Cell Margins the tick will disappear
C. Right-click the Cross-Tab and select Grid options and click edit cell margins and remove cell margins the tick will disappear
D. Right-click the Cross-Tab and select Cross-Tab Grid options and click remove cell margins the tick will disappear

Q25. Which of the following can you perform within the Customize Style Tab? *(Multiple Answers)*

A. Suppress Empty Columns
B. Suppress Empty Rows
C. Suppress Columns Grand Total
D. Suppress Row Grand Total
E. Format Grid lines
F. Repeat Row Labels
G. Display String Value

Q26. You want to show the Grid lines within the Cross-Tab. Which of the following apply?

A. Right-click the Cross-Tab and select Cross-Tab Expert, select the Customize Style Tab and click the Format Gridlines button and uncheck the Show Grid Lines checkbox
B. Right-click the Cross-Tab and select Cross-Tab Expert, select the Grid Options Tab and click the Format Gridlines button and uncheck the Show Grid Lines checkbox

C. Right-click the Cross-Tab and select Cross-Tab Expert, select the Customize Style Tab and check the Show Grid Lines checkbox

D. Right-click the Cross-Tab and select Cross-Tab Expert, select the Style Tab and click the Format Gridlines button and uncheck the Show Grid Lines checkbox

Q27. Totals appear on the top left section of the Cross-Tab. You want to suppress the Grand Total. Which of the following apply?

A. Right-click the Total text within the Cross-Tab and select Row Grand Totals and select Suppress Grand Totals

B. Right-click the Total text within the Cross-Tab and select Grand Totals and select Suppress

C. Right-click the Total text within the Cross-Tab and select Totals and select delete Grand Total

D. Right-click the Total text within the Cross-Tab and select move Grand Totals

Q28. You want to add a calculation to the Cross-Tab illustrated below to display the difference between the numbers of candidates who took the Crystal Reports 2008 exam in comparison to candidates who took the Crystal Reports 7 exam. What method can be applied to achieve this calculation?

	London	Birmingham	Manchester
Crystal Reports 2008	200000	40000	100000
Crystal Reports XI	7000	10000	430
Crystal Reports XI R2	400988	1000	722
Crystal Reports 10	20000	577	211
Crystal Reports 8.5	7000	9000	0
Crystal Reports 7	12000	100	76000

Fig: 7.2

A. Use a Calculated Member

B. Use a Group Option

C. Use a CurrentColumnIndex

D. Use a CurrentRowIndex

Q29. Which of the following Calculated Member formulas will produce the difference between the number of candidates who took the Crystal Reports 2008 exam and the number of Candidates that took the Crystal Reports 7 Exam

A. GridValueAt(GetRowPathIndexOf("Crystal Reports 2008"), CurrentColumnIndex, CurrentSummaryIndex) - GridValueAt(GetRowPathIndexOf("Crystal Reports 7"), CurrentColumnIndex, CurrentSummaryIndex)

B. GridValueAt(GetRowIndexOf("Crystal Reports 2008"), CurrentColumnIndex, CurrentSummaryIndex) - GridValueAt (GetRowIndexOf("Crystal Reports 7"), CurrentColumnIndex, CurrentSummaryIndex)

C. GridValueAt(GetRowOf("Crystal Reports 2008"), CurrentColumnIndex, CurrentSummaryIndex) - GridValueAt(GetRowOf("Crystal Reports 7"), CurrentColumnIndex, CurrentSummaryIndex)

Q30. Which of the following methods can be used to insert a Calculated Member formulas which will produce the difference between the number of candidates who took the Crystal Reports 2008 exam and the number of Candidates that took the Crystal Reports 7 Exam?

A. Right-Click the Crystal Reports 7 row and select Calculated Member – Select "Crystal Reports 7" as the First Value, right-click "Crystal Reports 2008" and select Difference of "Crystal Reports 2008" and "Crystal Reports 7", this will insert the New Calculated Member into the Cross-Tab grid

B. Right-Click the Crystal Reports 2008 row and select Calculated Member – Select "Crystal Reports 2008" as the First Value, right-click "Crystal Reports 7" and select Difference of "Crystal Reports 7" and "Crystal Reports 2008", this will insert the New Calculated Member into the Cross-Tab grid

Q31. What are the four main calculations available to Calculated Members? *(Multiple Answers)*

A. Difference
B. Quotient
C. Sum

D. Product of

E. Minus

Q32. Which of the following are Row Options within a Cross-Tab? *(Multiple Answers)*

A. Repeat Label on Page Break

B. Suppress Empty Rows

C. Indent Labels

D. Group Options

E. Row Group Sort Expert

Q33. You want to add a summary calculation to your existing Cross-Tab. Which of the following apply?

A. Right-click the Cross-Tab and select Advanced Calculations – Embedded summary – New – Edit Embedded summary and add your calculation – Save and Close and assign a name under description

B. Right-click the Cross-Tab and select Advanced Member Calculations – Embedded summary – New – Edit Embedded summary and add your calculation – Save and Close and assign a name under description

C. Right-click the Cross-Tab and select Member Calculations – Embedded summary – New – Edit Embedded summary and add your calculation – Save and Close and assign a name under description

Q34. You want to apply a TOP 2 limitation to your Cross-Tab, which of the following apply?

A. Right-click the Cross-Tab and select Group Record Sort Expert – Set Group sort to Top N, based on the calculated summary and set N = 2 and click ok

B. Right-click the Cross-Tab and select Group Sort Expert – Set for this Group sort to Top N, based on the calculated summary and where N = 2 and click ok

C. Right-click the Cross-Tab and select Record Sort Expert – Set Group sort to Top N, based on the calculated summary and set N = 2 and click ok

D. Right-click the Cross-Tab and select Sort Expert – Set Group sort to Top N, based on the calculated summary and set N = 2 and click ok

Q35. **Which of the following are new to Crystal Reports 2008 Cross-Tab creation?** *(Multiple Answers)*
A. Derived Rows
B. Embedded Summaries
C. Calculated Member
D. Show Grid Options
E. Derived Cells
F. Derived Columns

Q36. **Calculated Members do not show in the Cross-Tab Expert.**

A. True
B. False

Q37. **Embedded Summaries do not show in the Cross-Tab Expert.**

A. True
B. False

Q38. **You have inserted a calculated member displaying the difference between the number of Crystal Reports 2008 exam candidates and the Crystal Reports 7 candidates, the description appears as "difference", how can this be changed to a more logical description?**

A. Right-click the Cross-Tab – Advanced Calculations – Calculated Member – Click Edit Header Formula and Enter the required description
B. Right-click the Cross-Tab – Advanced Calculations – Calculated Member – Click Edit Row Value Formula and Enter the required description
C. Right-click the Cross-Tab – Advanced Calculations – Calculated Member – Click Edit Insertion Formula and Enter the required description

Q39 You have inserted a Calculated Member within your Cross-Tab, to highlight the difference between the number of candidates who took the Crystal Reports 2008 and Crystal Reports 7 exam, the calculated member appears under Crystal Reports 7 as illustrated in the diagram below, how can you move it to appear under Crystal Reports 2008?

	London	Birmingham	Manchester
Crystal Reports 2008	200000	40000	100000
Crystal Reports XI	7000	10000	430
Crystal Reports XI R2	400988	1000	722
Crystal Reports 10	20000	577	211
Crystal Reports 8.5	7000	9000	0
Crystal Reports 7	12000	100	76000
Difference	188000	39900	24000

A. Right-Click the calculated member and select Calculated Member – Edit Header Formula and Change the GridRowColumnValue to **Crystal Reports 2008,** the formula should appear as follows: GetRowGroupIndexOf(CurrentRowIndex) = 1 and GridRowColumnValue("Product.Product Name") = "**Crystal Reports 2008**"

B. Right-Click the calculated member and select Calculated Member – Edit Row Value Formula and Change the GridRowColumnValue to **Crystal Reports 2008,** the formula should appear as follows: GetRowGroupIndexOf(CurrentRowIndex) = 1 and GridRowColumnValue("Product.Product Name") = "**Crystal Reports 2008**"

C. Right-Click the calculated member and select Calculated Member – Edit Insertion Formula and Change the GridRowColumnValue to **Crystal Reports 2008,** the formula should appear as follows: GetRowGroupIndexOf(CurrentRowIndex) = 1 and GridRowColumnValue("Product.Product Name") = "**Crystal Reports 2008**"

Q40. Which of the following conditional formatting will turn the value in the cross-tab to red if the customer's region equals Warsaw?

A. if GridValueAt ("Customer.Region") = "Warsaw" then crred else crblack

B. if GridLabelAt ("Customer.Region") = "Warsaw" then crred else crblack

C. if GridRowColumnValue ("Customer.Region") = "Warsaw" then crred else crblack

D. if GridColumnRowValue ("Customer.Region") = "Warsaw" then crred else crblack

Use the Running Total Expert

Q1. Which of the following precedes a Running Total?

    A. ~
    B. #
    C. @
    D. ?

Q2. You have the following data as illustrated in fig 5.1, you have created a Running Total with the following settings:
=

    A. Field to summarize = Sales Amount
    B. Type of summary = Sum
    C. Evaluate for each record
    D. Reset =Never.

SAMPLE DATA - *fig 5.1*

Sales Representative	Sales Amount	Sales Date	Clients City
Antonia Iroko	136.47	10/07/2009	London
Antonia Iroko	185.20	11/08/2009	London
Antonia Iroko	68.00	12/08/2009	New York

What will the running total display in the report?

A

Sales Representative	Sales Amount	Sales Date	Clients City
Antonia Iroko	136.47	10/07/2009	London
Antonia Iroko	321.67	11/08/2009	London
Antonia Iroko	389.67	12/08/2009	New York

B

Sales Representative	Sales Amount	Sales Date	Clients City
Antonia Iroko	136.47	10/07/2009	London
Antonia Iroko	185.20	11/08/2009	London
Antonia Iroko	389.67	12/08/2009	New York

Q3. Using the sample data in fig 5.1, you have created a Running Total with the following settings: =

A. **Field to summarize** = Sales Amount
B. **Type of summary** = Sum
C. **Evaluate** = for each record
D. **Reset** =On change of Field = Client City.

A

Sales Representative	Sales Amount	Sales Date	Clients City
Antonia Iroko	136.47	10/07/2009	London
Antonia Iroko	321.67	11/08/2009	London
Antonia Iroko	389.67	12/08/2009	New York

B

Sales Representative	Sales Amount	Sales Date	Clients City
Antonia Iroko	136.47	10/07/2009	London
Antonia Iroko	204.47	11/08/2009	London
Antonia Iroko	68.00	12/08/2009	New York

Q4. Which of the following methods can be used to create a Running Total?

A. From the menu bar select insert Running Totals Field within the Field Explorer – Select New – Assign a Running Total Name – Select the Field to summarize and the summary type – Set the evaluate required and the reset required and click ok

B. Right – click the Running Totals Field within the Field Explorer – Select New – Assign a name within the Running Total Name box – Select the Field to summarize from the available tables and fields and use the arrow to transfer the field to Field to summarize, select summary type – Set the evaluate required and the reset required and click ok

C. From the menu bar select view Running Totals Field within the Field Explorer – Select New – Assign a Running Total Name – Select the Field to summarize and the summary type – Set the evaluate required and the reset required and click ok

D. From the menu bar select File Running Totals Field within the Field Explorer – Select New – Assign a Running Total Name – Select the Field to summarize and the summary type – Set the evaluate required and the reset required and click ok

Q5. **Under Evaluate within the Running Total dialogue box which of the following are available?** *(Multiple Answers)*

A. For each record
B. On change of field
C. On change of group
D. Use Group
E. Use formula

Q6. **Under Reset within the Running Total dialogue box; which of the following are available?** *(Multiple Answers)*

A. Always
B. Never
C. On change of field
D. On change of group
E. Use Group
F. Use formula

Q7. **Using fig 5.1, you have placed the Running Total field in the Report Header; with the following settings =**

Field to summarize = Sales Amount
Type of summary = Sum
Evaluate for each record
Reset =Never.
Which of the following results is applicable?

SAMPLE DATA - *fig 5.1*

Sales Representative	Sales Amount	Sales Date	Clients City
Antonia Iroko	136.47	10/07/2009	London
Antonia Iroko	185.20	11/08/2009	London
Antonia Iroko	68.00	12/08/2009	New York

A. 136.47
B. 185.20
C. 68.00
D. 389.67

Q8. Using fig 5.1, you have placed the **Running Total field in the Report Header; with the following settings =**
Field to summarize = Sales Amount
Type of summary = Sum
Evaluate for each record
Reset =Never.

Which of the following results is applicable?

SAMPLE DATA - *fig 5.1*

Sales Representative	Sales Amount	Sales Date	Clients City
Antonia Iroko	136.47	10/07/2009	London
Antonia Iroko	185.20	11/08/2009	London
Antonia Iroko	68.00	12/08/2009	New York

    A. 136.47
    B. 185.20
    C. 68.00
    D. 389.67

Q9. Using fig 5.1, you have placed the **Running Total field in the Page Header with the following settings =**

Field to summarize = Sales Amount
Type of summary = Sum
Evaluate for each record
Reset =Never.

Which of the following results is applicable?

SAMPLE DATA - *fig 5.1*

Sales Representative	Sales Amount	Sales Date	Clients City
Antonia Iroko	136.47	10/07/2009	London
Antonia Iroko	185.20	11/08/2009	London
Antonia Iroko	68.00	12/08/2009	New York

    A. 136.47
    B. 185.20
    C. 68.00
    D. 389.67

Q10. **Using fig 5.1, you have placed the Running Total field in the Page Footer with the following settings =**

Field to summarize = Sales Amount
Type of summary = Sum
Evaluate for each record
Reset =Never.
Which of the following results is applicable?

    A. 136.47
    B. 185.20
    C. 68.00
    D. 389.67

Q11. **Using fig 5.1, you have placed the Running Total field in the Details Section with the following settings =**

Field to summarize = Sales Amount
Type of summary = Sum
Evaluate for each record
Reset =Never.
Which of the following results is applicable? *(Multiple Answers)*

SAMPLE DATA - *fig 5.1*

Sales Representative	Sales Amount	Sales Date	Clients City
Antonia Iroko	136.47	10/07/2009	London
Antonia Iroko	185.20	11/08/2009	London
Antonia Iroko	68.00	12/08/2009	New York

    A. 136.47
    B. 185.20
    C. 68.00
    D. 389.67
    E. 321.67

Q12. **Using fig 5.2, you have created a group based on Sales Representative and placed the Running Total field in the Details Section with the following settings =**

Field to summarize = Sales Amount
Type of summary = Sum
Evaluate for each record

Reset =Never.

Which of the following results is applicable? *(Multiple Answers)*
SAMPLE DATA - *fig 5.2*

Sales Representative	Sales Amount	Sales Date	Clients City
Antonia Iroko	136.47	10/07/2009	London
Antonia Iroko	185.20	11/08/2009	London
Antonia Iroko	68.00	12/08/2009	New York
John Robertson	129.31	10/09/2009	London
John Robertson	115.50	11/10/2009	London

    A. 136.47
    B. 185.20
    C. 68.00
    D. 389.67
    E. 321.67
    F. 518.98
    G. 634.48

Build a Report with Alerts

Q1.   **Formulas cannot be used within alerts**

    A. True
    B. False

Q2.   **You want to create an Alert. Which of the following must be completed for the Alert to work?**

    A. Choose Report |Select Alert, create or modify alerts, new, assign a name, you must enter a message, a condition and a name, before the alert will work
    B. Choose Report |Select Alert |create or modify alerts | new | assign a name | you do not have to enter a message, but you must assign a condition and a name, before the alert will work
    C. Choose Report |Select Alert | create or modify alerts| new | assign a name | you must enter a message, but a condition is not required, before the alert will work
    D. Choose Report |Select Alert | create or modify alerts| new | assign a name | you must enter a message, not name required and no condition is required, before the alert will work

Q3.   **You want to display the Alert message in the Report Header. Which of the following formulas apply?** *(Multiple Answers)*

Alert Name: SalesAlert
Alert Message: 'Sales Target for Current Week has been achieved'

    A. If IsAlertEnabled ("SalesAlert") = true then AlertMessage("SalesAlert")
    B. If IsAlertTriggered ("SalesAlert") = true then AlertMessage("SalesAlert")
    C. If AlertEnabled ("SalesAlert") = true then AlertMessage("SalesAlert")
    D. If AlertTriggered ("SalesAlert") = true then AlertMessage("SalesAlert")

Q4. **You want to delete the Property Alert. Which of the following apply?**

A. Select Format from the menu – Alerts – Create or Modify Alerts, highlight the Alert required and click the Delete button
B. Select View from the menu – Alerts – Create or Modify Alerts, highlight the Alert required and click the Delete button
C. Select Report from the menu – Alerts – Create or Modify Alerts, highlight the Alert required and click the Delete button
D. Select File from the menu – Alerts – Create or Modify Alerts, highlight the Alert required and click the Delete button

Q5. **You want to change the name of the Alert. Which of the following apply?**

A. Select Format from the menu – Alerts – Create or Modify Alerts, highlight the Alert required and click Edit and modify name as required
B. Select View from the menu – Alerts – Create or Modify Alerts, highlight the Alert required and click Edit and modify name as required
C. Select Report from the menu – Alerts – Create or Modify Alerts, highlight the Alert required and click Edit and modify name as required
D. Select File from the menu – Alerts – Create or Modify Alerts, highlight the Alert required and click Edit and modify name as required

Q6. **Which of the following will happen when you refresh a report with an Alert?**

A. The Report Alert dialog box will appear, click the Alert button to view records applicable to the Alert condition
B. The Report Alert dialog box will appear, click the View button to view records applicable to the Alert condition
C. The Report Alert dialog box will appear, click the Records button to view records applicable to the Alert condition
D. The Report Alert dialog box will appear, click the View Records button to view records applicable to the Alert condition

Q7. Alerts are activated when reports are refreshed

A. True
B. False

Q8. Which of the following are alert functions? *(Multiple Answers)*

A. AlertMessage()
B. IsAlertEnabled()
C. IsAlertTriggered()

Q9. Alerts are activated when reports are refreshed

A. True
B. False

Q10. You want to create a report which alerts users when a sales target has been met for the week? The Sales Target being £8000. Which of the following methods will achieve the results required?

A. Insert a group based on the Sales date, ensure the section is printed for each week, insert a summary based on the sales amount for the group, from the menu bar select Reports – Alerts – Create or Modify Alerts – New, enter the alert name – message: if Sum ({Sales.Amount}, {Sales.Date}, "weekly") > 8000 then 'Sales Target Achieved' and set the condition to Sum ({Sales.Amount}, {Sales. Date}, "weekly") > 8000

B. Insert a group based on the Sales date, ensure the section is printed for each week, insert a summary based on the sales amount for the group, from the menu bar select Alerts – Create or Modify Alerts – New, enter the alert name – message: if Sum ({Sales.Amount}, {Sales.Date}, "weekly") > 8000 then 'Sales Target Achieved' and set the condition to Sum ({Sales.Amount}, {Sales.Date}, "weekly") > 8000

C. Insert a group based on the Sales date, ensure the section is printed for each week, insert a summary based on the sales amount for the group, from the toolbar select File – Alerts – Create or Modify Alerts – New, enter the alert name – message: if Sum ({Sales.

Amount}, {Sales.Date}, "weekly") > 8000 then 'Sales Target Achieved' and set the condition to Sum ({Sales.Amount}, {Sales. Date}, "weekly") > 8000

D. Insert a group based on the Sales date, ensure the section is printed for each week, insert a summary based on the sales amount for the group, from the toolbar select Database - Reports – Alerts – Create or Modify Alerts – New, enter the alert name – message: if Sum ({Sales.Amount}, {Sales.Date}, "weekly") > 8000 then 'Sales Target Achieved' and set the condition to Sum ({Sales.Amount}, {Sales.Date}, "weekly") > 8000

Q11. You have created a report alert, however each time you refresh the report you are not prompted with an alert. Which explanation below applies?

A. Delete the alert and rebuild

B. The alert is corrupt, refresh the report again and save as, using a new name

C. You have based the alert on an invalid field, delete alert and rebuild

D. The alert has not been enabled – select report – alerts – create or modify alerts – edit and check the enabled checkbox

Q12. Your report alert is enabled, however each time you refresh the report you are not prompted with an alert. Which explanation below applies?

A. The alert is corrupt, refresh the report again and save as, using a new name

B. You have based the alert on an invalid field, delete alert and rebuild

C. Select File- Report Options – and check Display Alerts on Refresh

D. None of the above, once an alert is enabled, it will prompt the user when the report is refreshed

Build a Top N Report

Q1. **A Top N report can only be created when a group summary exist within the report?**

    A. True
    B. False

Q2. **Which of the following apply when creating a Top N?**

    A. Report Sort Expert
    B. Group Record Sort Expert
    C. Group Sort Expert
    D. Section Sort Expert

Q3. **You want to create a Top N (5) report to show the number of clients recruited by Sales Representatives each year; you have already created a summary count of clients per Sales Representative. Which of the following methods apply?**

    E. select Report – Report Sort Expert - Top N for this group sort – based on count of clients – where Top N equals 5
    F. select Report – Group Record Sort Expert - Top N for this group sort – based on count of clients – where Top N equals 5
    G. select Report – Group Sort Expert - Top N for this group sort – based on count of clients – where Top N equals 5
    H. select Report – Section Sort Expert - Top N for this group sort – based on count of clients – where Top N equals 5

Q4. **Within the Group Sort Expert which of the following Group Sorts are available?** *(Multiple Answers)*

    A. All
    B. Top N
    C. Bottom N
    D. Top Percentage N
    E. Top Percentage
    F. Bottom Percentage N
    G. Bottom Percentage

Q5. **Top N records with similar values can be included by ticking one of the following?**

A. Include Others
B. Include Ties
C. Include All
D. Include all Other Ties ·

Q6. **You want to create a flexible Top N report giving the user the functionality to set the Top N sales value per representative as required when running a report. Which of the following methods apply?**

A. Create a number parameter by right –clicking the parameter Fields within the Field Explorer and select New – within the create new parameter dialogue box Assign a parameter name under name – select Number as type –and static under list of value and click OK; from the menu bar select Report – Group Sort Expert Top N for this group sort – based on sales amount – where Top N equals click the formula button and enter the parameter field within the formula workshop

B. Create a number parameter by right –clicking the parameter Fields within the Field Explorer and select New – within the create new parameter dialogue box Assign a parameter name under name – select Number as type –and static under list of value and click OK; from the menu bar select Report – Record Sort Expert Top N for this group sort – based on sales amount – where Top N equals click the formula button and enter the parameter field within the formula workshop

Q7. **You want to create a Top N summary report, a summary must already exist in the report before a Top N report can be created?**

A. True
B. False

RDCR08301

CHAPTER 6 - USE REPORT PROCESSING TECHNIQUES

Chapter 6 covers questions on using report processing techniques, this includes the multi pass processing technique, using the database expert, configuring data sources, updating and validating reports, distributing personalised content, using XML transforms, linking and join types.

Keywords:
Pass processing, Database Expert, Configuration, Personalised Content, XML, Transform, Links, Joins, Auto Link.

Explain the multi-pass reporting process

Q1.   Which of the following are Crystal Reports report processing stages? *(Multiple Answers)*

A. Pass 1
B. Pass 2
C. Pre-Pass 1
D. Pre-Pass 2
E. Pre-Pass 3
F. Pass 4
G. Pass 3

Q2.   What report processing stage is applicable to Total Page Count'?

A. Pre-Pass 1
B. Pre –pass 2
C. Pass 3
D. Pass 1
E. Pass 4
F. Pre-pass 3
G. Pass 4

Q3. **One of the formulas below is processed under the Pre-Pass 1. Which one is applicable?**

A. If WeekdayName (DayOfWeek ({Orders.SalesDate})) = 'Wednesday' then 'Required' else 'Not Required'

B. 2 x 3

C. "Order Number " + totext ({Orders.Order ID}) + " is required in " + totext({Orders.Required Date} - {Orders.Order Date}) + "days"

D. WhilePrintingRecords; NumberVar array EHArray; Sum(EHArray [1 to 10])

Q4. **Recurring formulas are evaluated under which one of the following?**

A. Pre-Pass 1
B. Pre –pass 2
C. Pass 3
D. Pass 1
E. Pass 4
F. Pre-pass 3
G. Pass 4

Q5. **Constant formulas are processed under which of the following report processing stages?**

A. Pre-Pass 1
B. Pre –pass 2
C. Pass 3
D. Pass 4
E. Pre-pass 3
F. Pass 4

Q6. **Formulas which fall within the BeforeReadingRecords category are processed before records are read from the database.**

A. True
B. False

Q7. **Which of the following fall into the Pre-Pass 1 stage?**

A. WhilePrintingRecords
B. WhileReadingRecords
C. BeforeReadingRecords
D. EvaluateAfter

Q8. **Which of the following are processed under Pass 1?**

A. WhilePrintingRecords
B. WhileReadingRecords
C. BeforeReadingRecords
D. EvaluateAfter

Q9. **Which of the following report processing functions can be applied to the formula below?**

If {Client.City} = 'London' then '10% Discount Applicable' else '2% % Discount Applicable'

A. EvaluateBefore
B. EvaluateAfter
C. BeforeReadingRecords
D. WhileReadingRecords

Q10. **Which of the following report processing functions can be applied to the formula below?**

If Sum({Sales.Value},{Sales.Representative}) > 200000 then 'Top Sales Person'

A. WhilePrintingRecords
B. EvaluateAfter
C. BeforeReadingRecords
D. WhileReadingRecords

Q11. **A report with saved data is processed under which report processing stage?**

A. Pass 1
B. Pass 2

C. Pre-Pass 1
D. Pre-Pass 4
E. Pre-Pass 2

Q12. Running Totals are processed under which of the following?

A. Pre-Pass 1
B. Pre –pass 2
C. Pass 3
D. Pass 1
E. Pass 4
F. Pre-pass 3
G. Pass 2

Q13. 'TopN' and 'Bottom N' are processed under which of the following stages?

A. Pass 1
B. Pass 2
C. Pre-Pass 1
D. Pre-Pass 4
E. Pre-Pass 2

Q14. Which of the following are processed under Pass 2? (Multiple Answers)

A. Cross-Tabs
B. Subreports
C. Group Selection
D. Running Totals
E. WhilePrintingRecords
F. PrintTime Formula
G. Grand Total Summary
H. OLAP Grids
I. Charts
J. Page On Demand

Q15. You have created a formula and you want this particular formula to be processed after the result of another formula has been retrieved. Which of the following should you use?

A. WhileReadingRecords
B. WhilePrintingRecords
C. EvaluateAfter ()
D. BeforeReadingRecords

Q16. Sorting, Grouping and Totalling are process under which of the following stages?

A. Pre-Pass 1
B. Pre –pass 2
C. Pass 3
D. Pass 1
E. Pass 4
F. Pre-pass 3
G. Pass 2

Q17. You have used a running total to create a Cross-Tab, under which stage will it be processed?

A. Pre-Pass 1
B. Pre –pass 2
C. Pass 3
D. Pass 1
E. Pass 4
F. Pre-pass 3
G. Pass 2

Q18. The creation of pages on demand is processed under which of the following stages?

A. Pre-Pass 1
B. Pre –pass 2
C. Pass 3
D. Pass 1
E. Pass 4
F. Pre-pass 3
G. Pass 2

Q19. Hierarchical grouping takes place under which of the following stages?

A. Pre-Pass 1
B. Pre –pass 2
C. Pass 3
D. Pass 1
E. Pass 4
F. Pre-pass 3
G. Pass 2

Q20. OLAP Grids and Subreports are processed under which of the following stages?

A. Pre-Pass 1
B. Pre –pass 2
C. Pass 3
D. Pass 1
E. Pass 4
F. Pre-pass 3
G. Pass 2

Q21. Maps are processed under which of the following stages?

A. Pre-Pass 1
B. Pre –pass 2
C. Pass 3
D. Pass 1
E. Pass 4
F. Pre-pass 3
G. Pass 2

Use the Database Expert

Q1: Tables are linked within which section?

A. Database Expert - Links Tab
B. Database Expert report - table tab
C. Database Data - link table tab
D. Database data - table Links Tab

Q2: Viewing database properties involve which of the following methods?

A. Right-click the table within the Database Expert (Links Tab) and select properties from the dropdown menu
B. Right-click the database connection within the Database Expert (Data Tab) and select properties from the dropdown menu
C. Right-click the database connection within the Database Expert (Database Tab) and select properties from the dropdown menu
D. Right-click the table within the Database Expert (Linking Tab) and select properties from the dropdown menu

Q3: Which two of the following are Database Expert connectivity options? *(Multiple Answers)*

A. My Connections
B. Repository based SQL Command
C. Historical data sources
D. Create New Connection
E. Customized Favorites

Q4: How can you add your current database connection to your favourite's folder?

A. From the Menu Bar select file and save as into the favourites folder
B. From the Database Expert, drag and drop the database into the folder
C. Right-click the database within the Current Connections and select the add to favourites
D. From the Database Expert, select file and save as into the favourites folder
E. The favourites folder does not exist in Crystal Reports 2008

Q5: **You want to check your current connections which of the following apply?**

A. File- Database - Current Location
B. Report - Database Expert - Current Location
C. Database - Database Expert – My Connections
D. View - Database - Database Expert - Current Connections

Q6: **How can you sort field names in a database table within the Field Explorer alphabetically?**

A. Right-click the database from the File expert and select option from the drop-down menu, under list table and description, select the button for both.
B. Right-click the database from the view expert and select option from the drop-down menu, under list table and description, select the button for both.
C. Right-click the database from the report expert and select option from the drop-down menu, under list table and description, select the button for both.
D. Right-click the database table within the Field Explorer and select sort fields alphabetically.
E. Right-click the database from the SQL expert and select option from the drop-down menu, under list table and description, select the button for both.

Q7: **Which of the following activates the Map field dialog?** *(Multiple Answers)*

A. A field name has changed
B. A field name no longer exist
C. A field data type has changed
D. A field contains more data

Q8: **You want to view the SQL coding generated by your report. Which of the following apply?**

A. Choose File - Show SQL Query
B. Choose View - Show SQL Query
C. Choose Database - Database Expert -Show SQL Query

D. Choose Database - Show SQL Query

E. Choose Report - Show SQL Query

Q9:   You can edit the SQL generated by your report from within the Show SQL Query dialog box?

A. True

B. False

Q10:   You want to disconnect one of the two data sources you are connected to. Which of the following apply? *(Multiple Answers)*

A. Select Database from the Menu Bar and select Disconnect Database, from the Data Explorer, select the Database and click log off

B. Select Database from the Menu Bar and select disengage, from the Data Explorer, select the Database and click log off

C. Select Database from the Menu Bar and select End Session, from the Data Explorer, select the Database and click log off

D. From the Field Explorer, right-click the database and select Log on or off Server, from the Data Explorer highlight database and click log off and close

E. Select Database from the Menu Bar and select Log on or off Server, from the Data Explorer highlight database and click log off and close

Q11.   The database location can be changed within the Field Explorer.

A. True

B. False

Q12.   Which of the following are the two tabs available under the Database Expert?

A. Data

B. Links

C. Table

D. Field

Set Up and Explain how to Configure Data Sources

Q1: **You are connected to a SQL Server database via an ODBC connection. Which of the following database properties can you view within the Database Expert?** *(Multiple Answers)*

A. User ID
B. Password
C. Database
D. Database Type
E. Data Source Name
F. Use DSN Default Properties
G. File Name
H. DSN

Q2: **Which of the following describes the connection process to a Data source via an ODBC?**

A. File - New - Blank Report, expand the Create New Connection within the Database Connection dialog box - click the ODBC connection - double click the Make New Connection and select the Data source Name and click Next, enter your password and user is and click Finish, select the cross beside the tables and select the required tables using the > arrow to put reports in the selected tables section, click the Links Table and create the appropriate links

B. File, New, Blank Report, the Database Expert dialog box will appear, expand the Create New Connection within Available Data Sources and the cross beside the ODBC connection, double click the Make New Connection and select the Data source name and click Next, enter your password and user id and click Finish, select the required tables using the > arrow to put reports in the selected tables section, click the Links Table and create the appropriate links

Q3: **Set Datasource Location is used for which of the following?** *(Multiple Answers)*

A. To set a new database location
B. Automatically changes the database location
C. Changes the database drivers to the new database source chosen

Q4: **Which of the following are advantages of using ODBC?** *(Multiple Answers)*

A. Flexibility
B. Changes can be implemented to use the same ODBC connection to connect to different databases
C. None of the above

Q5: **What are the five layers used to access ODBC data sources**

A. Crystal Report layer \ ODBC translation layer | ODBC layer | dbms TRANSLATION LAYER | Database layer
B. Crystal Report layer \ ODBC translation layer | ODBC Format| dbms TRANSLATION LAYER | Database layer
C. Crystal Report layer \ ODBB translation layer | ODBC layer | dbms TRANSLATION LAYER | Database layer
D. Crystal Report layer \ ODBB translation layer | ODBC layer | dbbms TRANSLATION LAYER | Database layer

Q6: **One of the advantages of using ODBC connections is its flexibility to access a wide range of data from one section.**

A. True
B. False

Q7: **You want to connect to a client/server database. What are the three general usable methods of connection?**

A. ODBC/OLEE/SIMPLE ACCESS
B. NONE OF THE ABOVE
C. ODBC ONLY
D. ODBC/OLE DB/Direct DB drivers

Q8: **Which of the following are database connectivity technologies?** *(Multiple Answers)*

A. ODBC
B. Access database
C. OLE DB
D. SQL Database

Q9:   Describe an OLE database connection *(Multiple Answers)*

A. OLE DB enables the communication between the reporting system and SQL Server
B. OLE DB enables the communication between the reporting system and Microsoft Access
C. OLE DB enables the communication between the reporting system and Microsoft Excel
D. OLE DB enables the communication between the reporting system and DBMS

Q10:   To set up an ODBC datasource you must have the ODBC data driver installed on your machine.

A. True
B. False

Q11:   The database password has changed; you need to reconfigure the password via the ODBC driver to re-establish connection to your report. Which of the following apply? (Windows Environment)?

A. Program Files - Administrative Tools - Data Sources (ODBC) datasource - Systems tab - highlight the data source to be changed and click the configure button, click next, enter the new password - test datasource.
B. All Programs – Connect To - double-click Administrative Tools - Data Sources (ODBC) datasource - Systems tab, highlight the data source to be changed – configure - enter the new password - Test datasource.
C. Program Files, double-click Administrative Tools – Services - systems tab, and highlight the data source to be changed - configure - enter the new password – test datasource.
D. From the control panel, double-click Administrative Tools, - Data Sources (ODBC) datasource, select the systems tab, highlight the data source to be changed and click the configure button, click next, enter the new password, click next, next again and click the test the datasource button, then click OK.

Q12:   **You want to set up an ODBC data source to connect to Microsoft SQL Server, what should you do?**

A. Within the Control Panel double click the ODBC datasource icon - Systems Tab, click Add, select the ODBC Driver, select Microsoft SQL ServerDriver (*.mdb), click finish, enter the datasource name and description in the box that appears and click the select button to select the database, click ok.

B. Within the Control Panel double click Administrative Tools – select the ODBC datasource icon, select the Systems Tab, click Add, select the ODBC Driver, select Microsoft SQL ServerDriver, click finish, enter the datasource name and description in the box that appears and click the select button to select the Server you would like to connect to, click ok, enter the authenticity of the login ID, change the default database and click NEXT and finish, your ODBC connection will now appear in the Database Expert when creating a Crystal Report.

C. Within the Control Panel click the ODBC datasource icon, select Systems Tab, click Add, select the ODBC Driver, select Microsoft SQL ServerDriver (*.mddb), click finish, enter the datasource name and description in the box and click ok.

D. Within the Control Panel double click the ODBC datasource icon, select the Systems Tab, click Add, select the ODBC Driver, select Microsoft SQL ServerDriver (*.mdbb), click finish, enter the datasource name and description in the box that appears and click the select button to select the database, click ok.

Q13:   **You want to set up an ODBC data source for Microsoft Access. Which of the methods would you employ?**

A. From the Control Panel double click the ODBC datasource icon, select the Systems Tab, click Add, select the ODBC Driver, select Microsoft Access Driver (*.mcdb), click finish, enter the datasource name and description in the box that appears and click the select button to select the database, click ok, your ODBC connection will now appear in the Database Expert when creating a Crystal Report.

B. From the Control Panel double click the ODBC datasource icon, select the Systems Tab, click Add, select the ODBC Driver, select Microsoft Access Driver (*.mdb), click finish, enter the datasource name and description in the box that appears and click the select button to select the database, click ok, your ODBC connection will now appear in the Database Expert when creating a Crystal Report.

C. From the Control Panel double click the ODBC datasource icon, select the Systems Tab, click Add, select the ODBC Driver, select Microsoft Access Driver (*.mddb), click finish, enter the datasource name and description in the box that appears and click the select button to select the database, click ok, your ODBC connection will now appear in the Database Expert when creating a Crystal Report.

D. From the Control Panel double click the ODBC datasource icon, select the Systems Tab, click Add, select the ODBC Driver, select Microsoft Access Driver (*.mdbb), click finish, enter the datasource name and description in the box that appears and click the select button to select the database, click ok, your ODBC connection will now appear in the Database Expert when creating a Crystal Report.

Q14: **You want to connect to a SQL Server database via an OLE connection. Which of the following methods would you employ?**

A. From the Database Expert, Select the Create New Connection, select ODBC (ADO), from the OLE DB (ADO) dialog box select Microsoft OLE DB Provider for SQL Server and click Next, enter the server, user ID, password and database and click next, the Advanced Information dialog box will appear, add or remove properties if required and click finish

B. From the Database Expert, Select the Create New Connection, select OLE DB file (DSN), from the OLE DB (ADO) dialog box select Microsoft OLE DB Provider for SQL Server and click Next, enter the server, user ID, password and database and click next, the Advanced Information dialog box will appear, add or remove properties if required and click finish

C. From the Database Expert, Select the Create New Connection, select OLE DB (ADO), from the OLE DB (ADO) dialog box select Microsoft OLE DB Provider for SQL Server and click Next, enter the server, user ID, password and database and click next, the Advanced Information dialog box will appear, add or remove properties if required and click finish

Update Reports for Database Changes

Q1: **You want to point the sales report database to a new location via the ODBC connection. Which of the following apply?**

A. Database - Set database location – highlight table within the Current Data Source section - within the Replace With section expand Create New Connection, expand ODBC (RDO) and Double Click Make New Connection, select New ODBC connection, enter user id and password, expand tables within the new connection, highlight table and click the update button

B. From the Menu Bar select File - database - set database location - highlight the table in the Current Data Source section of the Set Datasource Location dialog box, in the Replace With section click the cross beside Create New Connection, click the cross beside ODBC (RDO) and Double Click Make New Connection, select New ODBC connection, enter user id and password, expand Tables within the new connection, highlight table and click the update button

C. From the Menu Bar select View -database - set database location - highlight the table in the Current Data Source section of the Set Datasource Location dialog box, in the Replace With section click the cross beside Create New Connection, click the cross beside ODBC (RDO) and Double Click Make New Connection, select New ODBC connection, enter user id and password, expand Tables within the new connection, highlight table and click the update button

Q2: **You have set your report to a new location, the region_ name field, which you have not used in the current report, has been deleted from the new datasource and the EmployeeID field which you have used has been renamed as WorkerID, how would you proceed?**

A. The Map Fields dialog box will appear, in the unmapped fields section the Region_Code and employeeID will appear, uncheck the Match Type checkbox and highlight the employeeID in the report fields section and the WorkerID in the new datasource section and click the map button.

B. Choose - report - set location - highlight the current database in the current datasource section and highlight the new datasource in the replace with section and click the update button, the mapping dialogue box will appear, highlight the employeeID in the unmapped section and the worker_id in the mapped section and click on the map button, this will create a map of the employeeID as Worker_ID

C. Choose - File - set location - highlight the current database in the current datasource section and highlight the new datasource in the replace with section and click the update button, the mapping dialogue box will appear, highlight the employeeID in the unmapped section and the worker_id in the mapped section and click on the map button, this will create a map of the employeeID as Worker_ID

D. Choose - view - set location - highlight the current database in the current datasource section and highlight the new datasource in the replace with section and click the update button, the mapping dialogue box will appear, highlight the employeeID in the unmapped section and the worker_id in the mapped section and click on the map button, this will create a map of the employeeID as Worker_ID

Explain How to Validate Report Data and Validate Report Data

Q1: **How can you ensure your report is using the latest version of the database?**

    A. Refresh Report
    B. Close and open report
    C. Preview Report
    D. Verify database

Q2: **Which of the following are methods of distributing a Crystal Report?** *(Multiple Answers)*

    A. Faxing
    B. Exporting
    C. Via the Web
    D. Printing
    E. Via Business Objects Enterprise
    F. Via Business Objects Enterprise ONLY

Distribute Personalized Content

Q1: **What is the definition of DRDPs?**

    A. Distribution Recipient Data Providers (DRDPs)
    B. Data Recipient Data Providers (DRDPs)
    C. Dynamic Recipient Data Providers (DRDPs)
    D. Detailed Recipient Data Providers (DRDPs)

Q2: **Which of the following methods can be used to publish reports?** *(Multiple Answers)*

    A. Info View
    B. Business Objects Enterprise
    C. Email
    D. FTP

Q3: **Which of the following can be used to distribute reports to recipients who do not have Business Objects Enterprise accounts?**

    A. Distribution Recipient Data Providers (DRDPs)
    B. Data Recipient Data Providers (DRDPs)
    C. Dynamic Recipient Data Providers (DRDPs)
    D. Detailed Recipient Data Providers (DRDPs)

Q4: **Which of the following allows the application of report distribution criteria?**

    A. Delivery Rules
    B. Publish Rules
    C. Report Rules
    D. Export Rules

Q5: **You have created 2 DRDP's for one publication, which statement below applies?**

    A. Only one DRDP is allowed per publication
    B. only two DRDP are allowed per publication
    C. None of the above apply

Q6: **You want to distribute your report to a number of users who do not have Business Objects Enterprise accounts which of the following will aid this process?**

A. .Net file users information
B. Database with recipients information
C. None of the above

Q7: **Which of the following applies to DRDP's?**

A. .Net file users information
B. Database with recipients information
C. Universe queries
D. Desktop intelligence reports
E. Business Objects accounts
F. None of the above

Q8: **Which of the following processes occur when personalising a publication?** *(Multiple Answers)*

A. Report is created listing recipients details
B. Recipients details are mapped to publication
C. Business Objects account is created
D. None of the above

Use an Xml Transform

Q1: **XML Transform is associated with which of the following?**

A. XSLT document
B. XLT document
C. XMT document
D. XMMT document

Q2: **Which of the following best describes an XML export format?**

A. It is similar to an Microsoft Excel export
B. It can be transformed into an XLT document
C. It is mainly data orientated
D. It can be transformed into an XMMT document

Q3: **You want to export your file to an XML format. Which of the following apply?**

A. Report - Export - Export Report - XML Format - OK - from the XML Export Options select Crystal Reports XML (default) and click ok; save export to the required location.
B. Database - Export - Export Report - XML Format - OK - from the XML Export Options select Crystal Reports XML (default) and click ok; save export to the required location.
C. Edit - Export - Export Report - XML Format - OK - from the XML Export Options select Crystal Reports XML (default) and click ok; save export to the required location.
D. File - Export - Export Report - XML Format - OK - from the XML Export Options select Crystal Reports XML (default) and click ok; save export to the required location.

Q4: **You have exported your file into the default Crystal Reports XML format, you now want to convert the default output into a format of your choice, what is this document called?**

A. XLT transform document
B. XMT transform document

C. XMMT transform document

D. XSLT transform document

Q5:   Which of the following is a basic file used to create a transform file?

A. Notepad

B. MS Word

C. MS Excel

D. MS Access

Q6:   What is the definition of XML?

A. Extensible Markup Language

B. Extendable Markup Language

C. Extensions Markup Language

D. Extension Markup Language

Q7:   You want to delete an XML transform format you previously added to the report. Which of the following apply?

A. File - Export - XML Exporting Formats - select the format from the list and click delete and OK, save report for changes to take effect.

B. File - Export - Manage XML Exporting Formats - select the format from the list and click delete and OK, save report for changes to take effect.

C. File - Export – Manage Exporting Formats - select the format from the list and click delete and OK, save report for changes to take effect.

D. File - Manage XML Exporting Formats - select the format from the list and click delete and OK, save report for changes to take effect.

Q8:   Which of the following statements are applicable to XML? *(Multiple Answers)*

A. Open to publication on various platforms

B. Applicable for various language storage

C. Content conversion is highly applicable

D. Contains ASCII characters

Q9:   What does XML use to store formatting information?

A. HTML sheets
B. XLS sheets
C. HTM sheets
D. Style sheets

Q10:   How many tags does an element contain?

A. Two
B. One
C. Three
D. Four

Q11:   What is the role of elements in XML?

A. Acts as a Tag
B. Remove non ascii characters
C. Provide indentations
D. Give details of data

**Q12:   Which of the following does the XML Schema define?
Multiple Answers)**

A. Content
B. Interpretation
C. Structure
D. File Number

**Q13:   Which of the following are contained in an XML document?
*(Multiple Answers)***

A. Elements
B. Attributes
C. Reports
D. Roles

**Q14:   You want to Import an XML Transform.   Which method
applies?**

A. File – Manage XML Export Formats - click Add – from the Add a New XML Format dialog box – assign a name for the format - Click Import XML Transform - Browse to locate XSLT and select the file – enter the file type in the specify the file type text area and enter a description if required and click ok

B. File – Export - XML - click Add – from the Add a New XML Format dialog box – assign a name for the format - Click Import XML Transform - Browse to locate XSLT and select the file – enter the file type in the specify the file type text area and enter a description if required and click ok

C. File – Export – Manage XML Export Formats - click Add – from the Add a New XML Format dialog box – assign a name for the format - Click Import XML Transform - Browse to locate XSLT and select the file – enter the file type in the specify the file type text area and enter a description if required and click ok

D. Export – Manage XML Export Formats - click Add – from the Add a New XML Format dialog box – assign a name for the format - Click Import XML Transform - Browse to locate XSLT and select the file – enter the file type in the specify the file type text area and enter a description if required and click ok

Q15: You are creating a Transform document and you would like all image files to be exported, what can you do to ensure this happens?

A. Tick include images and binary data fields (BLOBS)
B. Tick include images ONLY
C. Tick include images files
D. Tick include binary data fields (BLOBS)

Q16: Which of the following are available datasources when a XML driver is selected? *(Multiple Answers)*

A. Local datasource
B. HTTP(s)
C. Web Services

Q17. Which of the following describes a local data source connection file?

A. XML File

B. RPT

C. UNV

Q18:   What is the definition of WSDL?

A. Web Service data warehouse

B. Web Service data

C. Web Services Description Language

Q19:   Style sheets with which of the following extensions work with XML?

A. XLS

B. XSL

C. LXL

Identify Links and Join Types

Q1: **The Links Tab within the Database Expert provides you with which of the following?**

A. Current tables, links and indexes, Auto Arrange only
B. Current tables, links and indexes, Auto Arrange, Auto-link, Order links only
C. Current tables select for report creation, their links and indexes, Auto Arrange, Auto-link, Order links, clear links only
D. Current tables used for report creation, links and index legend, Auto Arrange, Auto-link Order links, clear links, delete links and link options
E. Current tables, links and indexes, Auto Arrange, Auto-link, Order links, clear links, delete links and link options, update link

Q2: **Which of the following are methods of Auto-Links?** *(Multiple Answers)*

A. By Key
B. By Name
C. By Field
D. By Data Type

Q3: **The Database Expert Links Tab only appears if two or more tables are selected for report design?**

A. True
B. False

Q4: **You want to change the join type between two tables. Which methods apply?** *(Multiple Answers)*

A. Within the Database Expert Links Tab, double click the link between the two tables and change the join type
B. Within the Database Expert Links Tab, right-click the link between the two tables and choose link options from the drop down menu and change the join type
C. Within the Database Expert Links Tab, highlight the link between the two tables and click link options button

D. From the toolbar select link type select tables, right-click table and select link

Q5: **You right click the link between two tables in the Database Expert. Which of the following Link Options will be available?** *(Multiple Answers)*

A. Delete Link
B. Reconnect Link
C. Remove all Link
D. Reverse Link

Q6: **Which of the following removes all links created between tables in the Database Expert?**

A. Override Linking
B. Revert Linking
C. Clear Links
D. Remove All Links
E. Delete Links

Q7: **You try to link two tables together based on a field with different data types. Which of the following error messages will be displayed?**

A. Database is unreadable
B. You do not have permission to the database
C. You do not have permission to create this link
D. Data types are not compatible

Q8: **Which of the following buttons allows the Designer to activate the processing order for links?**

A. Arrange Links
B. Order Links
C. Auto Arrange Links
D. Smart Linking

Q9: You try to create smart linking within the Database Expert the link cannot be created. Which of the following could be the possible cause? *(Multiple Answers)*

A. Not supported by Database Driver
B. Database not connected
C. Driver Not Installed
D. Key relationships are not present

Q10: Which of the following is an essential aspect of database creation?

A. Schema relation Analysis
B. Normalization
C. Optimization relationship join
D. Concept Analysis

Q11: Which of the following link types are available in the Database Expert Link Tab? *(Multiple Answers)*

A. (=) link
B. [<] link
C. [<=] link
D. [! =] link
E. [=!] link
F. [>!]
G. [>] link
H. [>=] link

Q12: A rep_id link has been created between the Sales Representatives table and the sales table; you want to extract all representatives' sales. Which link type should you apply?

A. (=) link
B. [>] link
C. [>=] link
D. [<] link
E. [<=] link
F. [! =] link

Q13: You want to return records from the left table matching records from the right table every time the joining field in the left table is less than or equal to the joining field in the right table. Which one of the following links should you use?

A. (=) link
B. [>] link
C. [>=] link
D. [<] link
E. [<=] link

Q14: You want to return all combination of records from both tables where the joining fields are not equal. Which one of the following links should you use?

A. (=) link
B. [>] link
C. [>=] link
D. [<] link
E. [<=] link
F. [! =] link

Q15: You want to return records from the left table matching records from the right table every time the joining field in the left table is greater than or equal to the joining field in the right table. Which one of the following links should you use?

A. (=) link
B. [>] link
C. [>=] link
D. [<] link
E. [<=] link
F. [! =] link

Q16: You want to extract a comparison between Managers in the Accounts Department and Managers in the Manufacturing Department, you want to make sure Managers in the Accounts Department are not earning more than Managers in the Manufacturing Department; you have created a link based on the remuneration field. Which link type will you use?

A. (=) link
B. [>] link
C. [>=] link
D. [<] link
E. [<=] link
F. [! =] link

Q17: You want to return records from the left table matching records from the right table every time the joining field in the left table is less than the joining field in the right table. Which one of the following links should you use?

A. (=) link
B. [>] link
C. [>=] link
D. [<] link
E. [<=] link
F. [! =] link

Join Type

SAMPLE TABLES

Link both tables on location

ExamCode	Center Location
RDCR08201	Ascot
RDCR08301	London
RDCR08400	Brent

CandNo	Cand Location
CR1033	Purley
CR1099	London
CR5667	Brent

CRYSTAL_EXAM_CENTERS

CANDIDATE_LOC

Fig 3

Q18: Which of the following join types will produce the results as illustrated in the table below?

ExamCode	Center Location	CandNo	Cand Location
RDCR08201	Ascot	NULL	NULL
RDCR08301	London	CR1099	London
RDCR08400	Brent	CR5667	Brent

    A. Right Outer Join
    B. Full Join
    C. Left Outer Join
    D. Inner Join

Q19: Which of the following join types will produce the results as illustrated in the table below?

ExamCode	Center Location	CandNo	Cand Location
RDCR08301	Ascot	CR1099	Ascot
RDCR08400	Brent	CR5667	Brent

    A. Right Outer Join
    B. Full Join
    C. Inner Join
    D. Left Outer Join

Q20: You want to show all records in the CANDIDATE_LOC table that are unmatched to the records in the CRYSTAL_ EXAM_CENTERS. Which of the following will you use?

    A. Full Outer Join
    B. Left Outer Join
    C. Right Outer Join
    D. Inner Join

Q21: Using the tables in *Fig 3,* you link both tables on location and you have created a Left Outer Join between the tables. What will the result of this join be?

    A. Will include records in the right table (CRYSTAL_EXAM_ CENTERS) that are unmatched with rows in the left table (CANDIDATE_LOC)

B. Will include records in the left table (CRYSTAL_EXAM_CENTERS)
C. Will include records in the right table (CANDIDATE_LOC)
D. Will include records in the left table (CRYSTAL_EXAM_CENTERS) that are unmatched with rows in the right table (CANDIDATE_LOC)

Q22: Which of the following are join types in Crystal Reports?

A. Inner join, Left Outer Join, Right Outer Join, Full Outer Join, Equal Link
B. Inner join, Left Outer Join, Right Outer Join, Full Outer Join
C. Full Inner Join, Left Outer Join
D. Full Inner Join, Full Left Outer Join, Full Right Join

Join Type

SAMPLE TABLES

Link both tables on location

ExamCode	Center Location
RDCR08201	Ascot
RDCR08301	London
RDCR08400	Brent

CandNo	Cand Location
CR1033	Purley
CR1099	London
CR5667	Brent

CRYSTAL_EXAM_CENTERS   CANDIDATE_LOC

Fig 3

Q23: You want to retrieve all centers whether they have registered candidates or not. Which join type should you apply?

A. Inner Join
B. Left Outer Join,
C. Right Outer Join,
D. Full Outer Join

Q24: You want to retrieve all centers with registered candidates only. Which of the following should be used?

A. Full Outer Join
B. Inner Join

C. Left Outer Join

D. Right Outer Join

Q25: Which of the following join types will produce the results as illustrated in the table below?

A. Right Outer Join

B. Full Join

C. Left Outer Join

D. Inner Join

ExamCode	Center Location	CandNo	Cand Location
NULL	NULL	CR1033	Purley
RDCR08301	London	CR1099	London
RDCR08400	Brent	CR5667	Brent

Join Type

SAMPLE TABLES

Link both tables on location

ExamCode	Center Location
RDCR08201	Ascot
RDCR08301	London
RDCR08400	Brent

CandNo	Cand Location
CR1033	Purley
CR1099	London
CR5667	Brent

CRYSTAL_EXAM_CENTERS   CANDIDATE_LOC

Fig 3

Q26: Which of the following join types will produce the results as illustrated in the table below?

A. Right Outer Join

B. Full Join

C. Left Outer Join

D. Inner Join

ExamCode	Center Location	CandNo	Cand Location
NULL	NULL	CR1033	Purley
RDCR08301	London	CR1099	London
RDCR08400	Brent	CR5667	Brent
RDCR08201	Ascot	NULL	NULL

Q27: Which of the following are the Enforced Join types available via the Database Expert? *(Multiple Answers)*

A. Enforced but Not
B. Not Enforced
C. Enforced From
D. Enforced To
E. Enforced Both

Q28: Which Enforced Join type is the default option?

A. Enforced but Not
B. Not Enforced
C. Enforced From
D. Enforced To
E. Enforced Both

Q29: Which of the following links is enforced only when necessary?

A. Enforced but Not
B. Enforced From
C. Enforced To
D. Enforced Both
E. Not Enforced

Join Type

SAMPLE TABLES

Link both tables on location

ExamCode	Center Location
RDCR08201	Ascot
RDCR08301	London
RDCR08400	Brent

CandNo	Cand Location
CR1033	Purley
CR1099	London
CR5667	Brent

CRYSTAL_EXAM_CENTERS   CANDIDATE_LOC

Fig 3

Q30: You have created a link from the CRYSTAL_EXAM_
CENTERS table to the CANDIDATE_LOC table, you select
records from the CANDIDATE_LOC table and not from
the CANDIDATE_LOC table, and the select statement still
include records from the CRYSTAL_EXAM_CENTERS table.
Which of the following link options apply?

A. Enforced but Not
B. Not Enforced
C. Enforced From
D. Enforced To
E. Enforced Both

Q31: You select records from the CRYSTAL_EXAM_CENTERS
table but not from the CANDIDATE_LOC table. The select
statement still includes records from both tables. Which
of the following link options apply?

A. Enforced but Not
B. Not Enforced
C. Enforced From
D. Enforced To
E. Enforced Both

Process Data on the Server

Q1. **Which of the following reporting criteria works effectively with a server based grouping method?** *(Multiple Answers)*

A. Reports with groups
B. Specified groups ONLY
C. Reports with sort applied
D. Reports with hidden sections
E. TopN, Bottom, Average, Distinct Count, Sample Variance, Max, Min based reports
F. Summary Group Header or Group Footer based reports
G. Summaries not based on Group Header or Group Footer

Q2: **You want to perform grouping on the server and use indexes or server for speed which of the following apply?** *(Multiple Answers)*

A. Choose File - Report Options and check the use indexes or server for speed and check the Perform Grouping On Server checkbox
B. Choose File - Report and check the use indexes or server for speed and check the Perform Grouping On Server checkbox
C. Choose File - Report expert Options and check the use indexes or server for speed and check the Perform Grouping On Server checkbox
D. Choose Database - Database Expert - Right-click the database and select Options from the drop-down menu and check the use indexes or server for speed and check the Perform Grouping On Server checkbox

Q3: **What are the benefits of using the 'perform grouping on server' option?** *(Multiple Answers)*

A. Memory processing improvements
B. Transfer improvements
C. Connection time improvements
D. Improves processing speed

Q4: **Which of the following will improve report-processing performance?** *(Multiple Answers)*

A. Use indexes to link tables
B. Use indexes or server for Speed
C. Try to avoid the use of data type conversion
D. Use constant expressions
E. Use SQL expressions

Q5: **Which of the following refer to server based grouping?** *(Multiple Answers)*

A. The performance of reporting
B. Aggregate calculation based on Group Headers
C. No aggregate calculations
D. The performance of aggregate sectioning
E. Aggregate calculation on Group Footers of records
F. TopN calculations
G. Bottom N calculations

CHAPTER 7 - USE SUBREPORTS

The following chapter covers questions on Subreports and how they can be integrated into an existing report or created from the beginning; it also presents questions relating to the types of Subreports and their functionality. At the end of this chapter knowledge gained will be in the following sections, creating a Subreport, linking a Subreport, using Subreport parameters and formatting Subreports to display data from different data sources.

Keywords:
Subreport, Unlinked, Linked, On-Demand, Shared Variables, Un-linkable

Q1.   Which of the following best describes a Subreport?

A. Two reports created and saved as .rpt files
B. A main report and an embedded report
C. Two reports created from different data sources
D. None of the above
E. All of the above

Q2.   You are creating a Subreport from two different databases. Which of the following apply?

A. This cannot be done
B. Force a link between the two databases
C. Use a Global Variable
D. Create a Subreport with a shared variable
E. Create an Unlinked Subreport

Q3.   A shared variable from the main report can be passed to the Subreport via a parameter

A. True
B. False

Q4.   The records retrieved within a linked Subreport are based on the filter applied to the main Subreport?

A. True
B. False

Q5. **Which of the following are the two types of Subreports?**

A. Linked and Delinked Subreports
B. Linked and Unlinked Subreports
C. Equal and unequal Subreports
D. Main and Subreports
E. Main and unlinked Subreports

Q6. **What will appear when you double click a Subreport?**

A. Nothing will appear
B. Another Subreport
C. The Subreport design section will appear
D. A preview tab for the Subreport will appear

Q7. **You want to close the several tabs which appeared after you double-clicked your Subreport. Which of the following apply?**

A. Refresh the report
B. Preview the report
C. Close the report and open again
D. Click the cross which is situated next to the page navigation controls

Q8. **You preview a main report on its own preview tab; this will also result in the preview of the Subreport at the same time.**

A. True
B. False

Q9. **To pass data from the main report to the Subreport you can use which of the following?**

A. Shared Variable
B. Variable
C. NumberVar
D. StringVar
E. Global Variable

Q10. To activate a Subreport's Design Tab. Which of the following apply?

A. Preview Subreport
B. Right-click the Subreport and select edit Subreport
C. Double-click Subreport
D. Refresh main report

Q11. Your Subreport returns no data. Which of the following explains this non retrieval?

A. Format Subreport
B. Delete Subreport
C. Suppress conditionally or Suppress Blank Subreport has been applied
D. Re-import Subreport

Q12. You want your Subreport to appear as a Hyperlink with the clients name and year record details, which of the following apply?

A. Right click the Subreport and select Format Subreport from the drop-down menu, select the Common Tab and check the on-demand Subreport checkbox, click the x+2 button and paste the formula "Double Click to view " + GroupName ({Client.Client Name}) + " Client records for 2008 " into this section
B. Right click the Subreport and select Format Subreport from the drop-down menu, select the Subreport tab and check the Re-Import When Opening checkbox and click the x+2 button and paste the formula "Double Click to view " + GroupName ({Client.Client Name}) + " Client records for 2008 " into this section
C. Right click the Subreport and select Format Subreport from the drop-down menu, select the File tab and check the on-demand Subreport checkbox, click the x+2 button and paste the formula "Double Click to view " + GroupName ({Client.Client Name}) + " Client records for 2008 " into this section
D. Right click the Subreport and select Format Subreport from the drop-down menu, select the Subreport tab and check the On-Demand Subreport checkbox, click the x+2 formula button beside

On-Demand Subreport Caption and paste the following formula into this section "Double Click to view " + GroupName ({Client. Client Name}) + " Client records for 2008 "

Q13. What can you do to remove the border around a Subreport?

A. This border cannot be removed
B. If the border is removed the Subreport data will not show.
C. Right click the border and press delete
D. Right-click the Subreport and select Format Subreport, select the Border Tab and set all borders (left, right, top and bottom) to none

Q14. When you pass a parameter from the main report to the Subreport as a link, the Subreport will use the parameter in the Select Expert - Record Selection.

A. False
B. True

Q15. When previewing a Subreport on its own Preview Tab, the main report can also be viewed.

A. True
B. False

Q16. Which of the following will improve report processing of the main report when a Subreport is included? *(Multiple Answers)*

A. Run both reports separately
B. Separate both reports
C. Create an On-Demand Subreport
D. Debug the Subreport
E. Use indexed fields for linking the main report to the Subreport.

Q17. You want to save your Subreport as a separate report. What should you do?

A. From the Menu Bar, select report and save as

B. From the Menu Bar select report, options, Save Subreport As
C. Right click the Subreport and select Save Subreport As
D. A Subreport cannot be saved as a separate report

Q18.   Which of the following statements are true? *(Multiple Answers)*

A. You cannot Drill-Down into Subreports
B. You can Drill-Down into Subreports
C. A Subreport can be saved as a separate report
D. A Subreport can be created within a Subreport

Q19.   You roll the mouse over an on-demand Subreport what will happen?

A. The cursor disappears
B. The Subreport is refreshed
C. The cursor changes to the Standard Windows Hand and the name of the Subreport is displayed
D. The Subreport database path is shown

Q20.   The Re-Import option does not exist when you Right-click your Subreport. Which of the following could explain this missing option?

A. The main report should be refreshed
B. A link has not been established between the main report and the second Subreport
C. The first Subreport was imported into the main report and the second Subreport was created from scratch via the report wizard
D. None of the above

Q21.   You have created a Shared Variable for the Customer. Region sales value for 2008, which you will like to add to the 2009 data in your main report. Where should you place the reset formula within your main report?

A. Region Group Header
B. Details Section
C. Report Header
D. Page Header
E. Page Footer

Q22. You want to create a link based on the customer_id between your main report and Subreport, although the data held in both fields are the same, the customer id in the main report is a numeric field and the customer id in the Subreport is a text field. Which of the following methods should you use to create the link?

A. Create a formula to convert the customer_id to a number in the main report and use this formula as a link between the Subreport and main report
B. Create a formula to convert the customer_id to a number in the Subreport report and use this formula as a link between the Subreport and main report
C. Use the Field Explorer to create the following formula:- tovalue({customer_id}) and use this formula as a link
D. Use the Field Explorer to create the following formula:- Value({customer_id}) and use this formula as a link

Q23. Which of the following statements is true about Subreports? *(Multiple Answers)*

A. A formula can be used as a link between a Subreport and a main report
B. A Subreport cannot be linked to a main report via a formula
C. A parameter can be used as a link between the main report and the Subreport
D. Shared variables can be passed from a Subreport to a main report
E. Shared variables can be passed from main reports to Subreports

Q24. Which of the following statements is true? *(Multiple Answers)*

A. A Subreport includes its own layout
B. A Subreport Includes its own database connection
C. A Subreport Includes its own selection criteria
D. A Subreport Is always controlled by the main report
E. The unlinked Subreport does not communicate with the main report.
F. The unlinked Subreport communicates only at runtime
G. A linked Subreport is controlled by the main report

Q25. You want to change an existing link between the Subreport and the main report. Which of the following apply?

A. Choose Subreport, View, reports, Subreport links, or right click the Subreport and choose Change Subreport Links
B. Choose file, edit, Subreport links, or right click the Subreport and choose Change Subreport Links
C. Choose edit, Subreport links, or right click the Subreport and choose Change Subreport Links
D. Choose Format, Subreport links, or right click the Subreport and choose Change Subreport Links

Q26. Where can the Object Name of the Subreport be changed?

A. You cannot change the name of the Subreport, once it has been created
B. Right-click the Subreport and select change name
C. Right-click the Subreport and select Format Subreport, select the Common Tab and enter the name of the Subreport
D. Right-click the Subreport and select Format Subreport or select format and Format Subreport from the Menu Bar, click the Common Tab and type in the name of the Subreport object in the Object Name section of the tab.

Q27. You have created a report and will now like to include an unlinked Subreport. What should you do?

A. From the Menu Bar select File, Subreport, choose or create the Subreport, click OK and insert the Subreport in your main report.
B. From the Menu Bar select, insert Subreport, choose an existing report or create the Subreport using the Report Wizard click OK and insert the Subreport into the main report.
C. From the Menu Bar select view, Subreport, choose an existing report or create the Subreport, click OK and insert the Subreport in your main report.
D. From the Menu Bar select File, Report, Subreport, choose an existing report or create the Subreport, click OK and insert the Subreport in your main report.

Q28. You Drill-Down into the candidate name of the Subreport and the preview tab appears with the name if the candidate. How can you apply this format to your Subreport?

A. Right click the Subreport and select Format Subreport from the drop-down menu, select the Common Tab and check the on-demand Subreport checkbox, click the x+2 button and paste the formula GroupName ({Candidate.Name}) + " " + "Clients" into this section

B. Right click the Subreport and select Format Subreport from the drop-down menu, select the Border Tab and check the on-demand Subreport checkbox, click the x+2 button and paste the formula GroupName ({Candidate.Name}) + " " + "Clients" into this section

C. Right click the Subreport and select Format Subreport from the drop-down menu, select the Subreport tab and click the x+2 button beside the Subreport Preview Tab Caption and paste the formula GroupName ({Candidate.Name}) + " " + "Clients" into this section

D. Right click the Subreport and select Format Subreport from the drop-down menu, select the File tab and check the on-demand Subreport checkbox, click the x+2 button and paste the formula GroupName ({Candidate.Name}) + " " + "Clients" into this section

Q29. Which of the following variables are available to the Subreport from the main report?

A. Global
B. Shared Variable
C. Local
D. Shared Local

Q30. You have passed a variable from the main report to the On-Demand Subreport nothing happens. Why did this happen?

A. The shared variable is not functional because you failed to include a variable name

B. The shared variable is not functional because you failed to include an evaluation time

C. The shared variable is not functional because you failed to include it in the Subreport
D. An on-demand Subreport cannot share information with the main report as it does not process at the same time as the main report

Q31. **You have created a Shared Variable in the main report, which you would like to pass to the Subreport. Which of the following should be placed in the Subreport to display the results?**

WhilePrintingRecords;
Shared NumberVar ClientTotal:= Sum({Purchase.Amount});

A. WhilePrintingRecords; Shared NumberVar ClientTotal:= Sum({Purchase.Amount});
B. WhilePrintingRecords; Shared ClientTotal:= Sum({Purchase. Amount});
C. WhilePrintingRecords; NumberVar ClientTotal:= Sum({Purchase. Amount});
D. WhilePrintingRecords; Shared NumberVar ClientTotal

Q32. **A shared variable can be used to create a chart or a summary within the main report**

A. True
B. False

Q33. **A shared variable can be passed from an unlinked Subreport to a main report**

A. True
B. False

Q34. **You want to create a link between the main report and the Subreport; however you do not want the Subreport to use the link in its Record Selection. What should you do?**

A. This is not possible, the link must be used
B. Delete the Subreport and recreate the link
C. Un-check the select data in Subreport based on the field checkbox when creating the link between the main report and the Subreport

Q35. You want to create a linked Subreport. What should you do?

A. Choose, insert Subreport from the Menu Bar, choose an existing report or create one from scratch, click the link tab and select the field to be linked and click arrow > to select the field, choose the field from the main report to be linked to the field in the Subreport and click ok, places the Subreport within the main report as required

B. Choose, view, insert Subreport from the Menu Bar, choose an existing report or create one from scratch, click the link tab and select the field to be linked and click arrow > to select the field, choose the parameter field on the main report to be linked to the Subreport and choose the Subreport field, you can limit the size to records that match the main report linked field only.

C. Choose, report, insert Subreport from the Menu Bar, choose an existing report or create one from scratch, click the link tab and select the field to be linked and click arrow > to select the field, choose the parameter field on the main report to be linked to the Subreport and choose the Subreport field, you can limit the size to records that match the main report linked field only.

D. Choose, edit, insert Subreport from the Menu Bar, choose an existing report or create one from scratch, click the link tab and select the field to be linked and click arrow > to select the field, choose the parameter field on the main report to be linked to the Subreport and choose the Subreport field, you can limit the size to records that match the main report linked field only.

Q36. You place your cursor over the Subreport. What will appear?

A. Nothing will appear
B. Only the name of the Subreport will appear
C. The name of the Subreport and a magnifying glass will appear indicating the Drill-Down propensity of the report
D. The cursor will disappear

Q37. You have drilled down into a Subreport. What will happen?

A. Additional tabs will appear for each section drilled into

B. The Subreport will freeze and an error message will appear
C. The Subreport will request a parameter input.
D. None of the above

Q38. Several tabs appear after you have drilled down on a number of Subreports within your main report, what will happen when all tabs cannot be displayed on one page?

A. Two small left to right scroll arrows will appear on the right hand side of the group of tabs, enabling the user to scroll back and forward
B. You will be prompted with a limitation record message
C. The report will appear as a blank page
D. None of the above

Q39. You will like to re-import the Subreport each time it is opened. What should you do?

A. Right-click the Subreport and select Format Subreport tab, select the Subreport tab and check the Re-Import When Opening checkbox
B. Right-click the Subreport and select Format Subreport tab, select the Common Tab and check the Re-Import When Opening checkbox
C. Right-click the Subreport and select Format Subreport tab, select the Border Tab and check the Re-Import When Opening checkbox
D. Right-click the Subreport and select Format Subreport tab, select the font tab and check the Re-Import When Opening checkbox

Q40. You want to create a link between the main report and the Subreport, the field data type in the main report is a string and the field data type in the Subreport is a number. How can you achieve this link?

A. You cannot link the main report to the Subreport
B. Create a formula in the main report to change the field datatype to a number and create a link from the main report to the Subreport based on this formula
C. Change the datatype in the database
D. None of the above

Q41. You have created a Subreport, you notice only one line of data appears when the main report is refreshed; you know there is more data applicable to that Group. Which of the following should be implemented?

A. Right-click the Subreport and select the Subreport Can Grow Checkbox
B. Right-click the Subreport and select the Border Tab and check the Can Grow Checkbox
C. Right-click the Subreport and select the Report Tab and check the Can Grow Checkbox
D. Right-click the Subreport and select the Common Tab and check the Can Grow Checkbox

Q42. You are creating a Subreport using the Report Wizard. Which of the following sections will appear? *(Multiple Answers)*

A. Data, Fields
B. Grouping, Summaries
C. Formulas, Functions
D. Group Sorting, Record Selection
E. Charts, Template

Q43. Which of the following enhances the performance of Subreports? *(Multiple Answers)*

A. Use On-Demand Placeholder, processing only takes place when the user clicks the placeholder
B. Use indexed fields to link the main report to the Subreport
C. Always refresh the Subreport
D. None of the above

Q44. On-Demand Subreports appear as hyperlinks within the main report.

A. True
B. False

CHAPTER 8 - COMPLEX FORMULAS AND CUSTOM FUNCTIONS

Chapter 8 covers questions on creating complex formulas and custom functions, using evaluation time functions, using dynamic arrays, print state functions, loops and control structures and creating hyperlink reports.

Keywords:
Evaluation Time Functions, Print State Functions, Hyperlinks, Custom Functions, Control Structures, Loops, Arrays.

Use Evaluation Time Functions

Q1. **Which of the following are evaluation time functions? *(Multiple Answers)***

A. WhileReadingRecords
B. WhilePrintingRecords
C. EvaluateAfter ()
D. BeforeReadingRecords
E. AfterReadingRecords
F. DuringPrintingRecords

Q2. **Which of the following is user defined?**

A. BeforeReadingRecords
B. WhileReadingRecords
C. WhilePrintingRecords
D. EvaluateAfter
E. AfterReadingRecords

Q3. **You want the formula to be processed after formula b. Which of the following apply?**

A. BeforeReadingRecords
B. WhileReadingRecords
C. WhilePrintingRecords
D. EvaluateAfter

E. AfterReadingRecords
F. DuringPrintingRecords

Q4. What will the following formula produce?
BeforeReadingRecords;
{Customer.Contact First Name} + "   " + {Customer.Contact Last Name}

A. A concatenated field of First Name and Last Name
B. An Error Message: the field cannot be used because it must be evaluated later
C. None of the above

Q5. You are creating a manual running total. Which of the following are required? (Multiple Answers)

A. Reset Formula
B. Evaluate Formula
C. Display Result Formula
D. None of the above

Q6. Which of the following is an example of a reset formula?

A. WhileReadingRecords; Numbervar TotalValue:= 0
B. WhilePrintingRecords; Numbervar TotalValue:= 0
C. BeforeReadingRecords; Numbervar TotalValue:= 0
D. AfterReadingRecords; Numbervar TotalValue:= 0

Q7. Which of the following is an example of an evaluate formula?

A. WhileReadingRecords; Numbervar TotalValue:= TotalValue + {Orders Detail.Quantity}
B. WhilePrintingRecords; Numbervar TotalValue:= TotalValue + {Orders Detail.Quantity}
C. BeforeReadingRecords; Numbervar TotalValue:= TotalValue + {Orders Detail.Quantity}
D. AfterReadingRecords; Numbervar TotalValue:= TotalValue + {Orders Detail.Quantity}

Q8. **You want to create a Group running total, where will you place the reset formula?**

A. Report Header
B. Group Header
C. Group Footer
D. Report Footer
E. Details

Q9. **You want to create a Group running total, where will you place the evaluate formula?**

A. Report Header
B. Group Header
C. Group Footer
D. Report Footer
E. Details

Q10. **You want to create a Group running total, where will you place the result formula?**

A. Report Header
B. Group Header
C. Group Footer
D. Report Footer
E. Details

Also See: Multi-Pass Reporting Process

Use a Dynamic Array

Q1. **Which of the following have to be assigned when creating a Dynamic Array?** *(Multiple Answers)*

A. The Array must be declared
B. The array must be assigned a value
C. Elements in the array must be referenced
D. All of the above are optional

Q2. **A Dynamic Array is processed under which of the following passes?**

A. Pass 2
B. Pass 1
C. Pre-Pass 1
D. Pass 3

Q3. **You have declared a Dynamic Array. Which section of the report should this formula be placed?**

A. Group Header
B. Details
C. Report Footer
D. Page Footer

Q4. **You have assigned values to an array, where should this formula be placed?**

A. Group Header
B. Details
C. Report Footer
D. Page Footer

Q5. **The results formula of the Dynamic Array can be placed in which of the following sections?** *(Multiple Answers)*

A. Group Header
B. Details
C. Report Footer
D. Page Footer
E. Group Footer

Use Print State Functions

Record Number	First Name	Last Name	Exam Code	Score	Flag Nulls	Exam Date	Version
1	Antonia	Iroko	RDCR08200	77	77	05/06/2008	CR 2008
2	Antonia	Iroko	RDCR08600	58	58	02/02/2008	CR 2008
3	Antonia	Iroko	RDCR08500	50	50	09/07/2008	CR 2008
4	Antonia	Iroko	RDCR08300			08/08/2008	CR 2008
5	Antonia	Iroko	RDCR08400	89	89	08/01/2008	CR 2008
6	Antonia	Iroko	RDCR08100	80	80	07/06/2008	CR 2008
7	Martin	Lee	RDCR08100	79	79	19/06/2008	CR 2008
8	Chris	Harold	RDCR08500	100	100	05/06/2008	CR 2008
9	Chris	Harold	RDCR08100	85	85	05/07/2008	CR 2008
10	Chris	Harold	RDCR08200	66	66	12/08/2008	CR 2008
11	Chris	Harold	RDCR08300	88	88	05/09/2008	CR 2008
12	Mary	Miller	RDCR08100			13/12/2008	CR 2008
13	Paul	Hall	RDCR08100	60	60	31/08/2008	CR 2008
14	Philip	Johnson	RDCR08100			05/06/2008	CR 2008
15	Philip	Johnson	RDCR08200	79	79	05/09/2008	CR 2008
16	Philip	Johnson	RDCR08500	79	79	30/09/2008	CR 2008
17	Philip	Johnson	RDCR08300	89	89	17/12/2008	CR 2008
18	Philip	Johnson	RDCR08400	90	90	09/11/2008	CR 2008
19	Philip	Johnson	RDCR08100	46	46	31/10/2008	CR 2008
20	Tim	Goal	RDCR08100	90	90	05/06/2008	CR 2008

Q1. Based on the information in the table above, you have created a formula to flag nulls which exist in the score field; however the formula does not seem to work. Which of the following could be the possible cause?

if isnull({Sample_Exam_Results___2005.SCORE}) then 'Error' else {Sample_Exam_Results___2005.SCORE}

A. The Convert Database NULL Values to Default checkbox is checked
B. The Convert Database NULL Values to Default checkbox is unchecked
C. The convert other Null Values to default checkbox is unchecked
D. The convert other Null Values to default checkbox is checked

Q2. Which of the following functions can be used to handle null values in a formula? *(Multiple Answers)*

A. NextIsNull
B. IsNull
C. PreviousIsNull
D. Null

Q3. **Which of the following functions will flag the first null record?**

A. IsNull
B. NextIsNull
C. PreviousIsNull
D. Previous

Q4. **Which of the following Crystal functions will flag the last record?**

A. IsNull
B. NextIsNull
C. PreviousIsNull
D. Next

Q5. **You want to convert database values to default. What should you do?**

A. Choose Database\ Options and check the Convert Database NULL Values to Default and convert other Null Values to default
B. Choose Report\ Report Options and check the Convert Database NULL Values to Default and convert other Null Values to default
C. Choose Report and check the Convert Database NULL Values to Default and convert other Null Values to default
D. Choose File\ Report Options and check the Convert Database NULL Values to Default

Q6. **NextIsNull and PreviousIsNull cannot be used on databases that do not support Null Values**

A. True
B. False

Q7. **What is the definition of Null?** *(Multiple Answers)*

A. Zero in database field
B. No value in database field
C. Empty String
D. Spaces in database field
E. None of the above

Q8. **You count a null string field. It results in the following?**

A. Zero
B. Length of zero
C. Null
D. Error

Use Loop Control Structures with Arrays

Q1. Which of the following are While loop types? *(Multiple Answers)*

A. Do While
B. While
C. Do
D. While Do

Q2. Evaluates the condition first and if condition is met then it evaluates the expression after the Do; this process continues until the condition is false. Which of the following is being described?

A. Do While
B. While
C. Do
D. While Do

Q3. Evaluates the expression once, the condition is then evaluated second, the condition must be true before the expression is evaluated again; this process continues until the condition is false. Which of the following is being described?

A. Do While
B. While
C. Do
D. While Do

Q4. Which of the following enables the program to come out of a loop?

A. While Exit
B. Exit Only
C. Exit For
D. Exit While

Q5. **Which of the following are do Loop types?** *(Multiple Answers)*

A. Do While Loop
B. Do Until Loop
C. Do Loop While
D. Do Loop Until

Q6. **Which of the following evaluates the statement over and over again until the condition is true?**

A. Do While Loop
B. Do Until Loop
C. Do Loop While
D. Do Loop Until

Q7. **Which of the following are control structures?** *(Multiple Answers)*

A. If expressions
B. Select expressions
C. For Loops
D. While Loops
E. Option Loop
F. Loop Option

Also refer to Chapter 3 Create Formulas – Use Control Structures

Use Custom Functions

This chapter covers questions on Custom Functions; knowledge of Custom Functions is fundamental as it is a central function for creating and sharing customized formulas for present and future use hence reducing the processing time used to recreate formula. This chapter provides questions on the various functionalities of Custom Functions, their uses and limitations.

Q1. **You want to share your formula with other users. Which of the following will allow you to do this?**

   A. Operation Functions
   B. Report Functions
   C. Report Editor
   D. File Editor
   E. Custom Functions saved to the Repository

Q2. **Default Custom Functions are located in which one of the following?**

   A. Function Tree - Repository Custom Functions
   B. Formula Workshop - Repository Custom Functions
   C. Operations Tree - Repository Custom Functions
   D. Report Editor - Repository Custom Functions

Q3. **Which method will launch the Formula Workshop to create a Custom Function?**

   A. From the Menu Bar select Report | Formula Workshop Or click the Formula Workshop icon
   B. From the Menu Bar select File | Formula Workshop
   C. From the Menu Bar select insert Formula Workshop
   D. From the Menu Bar select Database Formula Workshop

Q4. **Which of the following represents the formula below when converted to a Custom Function?**

If {Student.ExamCenter} startswith 'York' then "Y001" + "\" + {Student.StudentID} else {Student.StudentID}

A. Function (StringVar e1, StringVar e2) If e1 startswith 'York' then "Y001" + "\" + Mv2 else Mv2

B. Func (StringVar v1, StringVar v2) if v1 startswith 'York' then "Y001" + "\" + v2 else v2

C. Function (StringVar v1, StringVar v2) if v1 startswith ' York ' then "Y001" + "\" + v2 else v2

D. Function (stVar v1, stVar v2) if v1 startswith ' York ' then "001" + "\" + v2 else v2

Q5. **Which one of the following is used to convert an existing formula to a Custom Function?**

A. Use Extractor

B. Use Editor

Q6. **You have moved an existing formula with 3 arguments to a Custom Function. What will the arguments be replaced with?**

A. B1,B2,B3

B. V1,V2.V3

C. Aa,Bb,Cc

D. Z1,Z2,Z3

Q7. **Which of the following statements apply to Custom Functions?** *(Multiple Answers)*

A. Custom Functions cannot use WhilePrintingRecords

B. Custom Functions cannot use any evaluation time functions

C. Custom Functions cannot use any summary functions

D. Custom Functions can only use local variable scope

Q8. **Syntax options in your Custom Function can be checked using which one of the following?**

A. Press Ctrl + A

B. Press Ctrl + C

C. Press Alt + C

Q9. **You have applied an evaluation time function to a Custom Function, this will not be allowed**

A. False
B. True

Q10. **Which one of the following is true?**

A. Custom Function names can contain spaces
B. Custom Function can start with numbers
C. The Custom Function name CBool is permissible
D. Names used by Crystal Report functions, cannot be used as Custom Function names

Q11. **Which of the following is used to create a new Custom Function?**

A. Use Extractor
B. Use Editor

Q12. **Custom Functions can be created using which of the following languages?** *(Multiple Answers)*

A. Crystal Syntax
B. Basic Syntax
C. Basic Syntax only
D. Crystal Syntax only

Q13. **Which of the following formulae can be converted into a Custom Function?** *(Multiple Answers)*

A. NUMBERVAR ArrayCounter;
STRINGVAR ARRAY Alphabet;
Alphabet:=MAKEArray("RDCR08201", "RDCR08301",
    "RDCR08401, "RDCR501");
ArrayCounter := ArrayCounter + 1;
Alphabet[ArrayCounter]
B.
Global Variable NumberVar ArrayCounter;
Global Variable StringVar array Alphabet;
Alphabet:=MAKEArray("RDCR08201", "RDCR08301",
    "RDCR08401, "RDCR501");

ArrayCounter := ArrayCounter + 1;
Alphabet[ArrayCounter]
C.
Local NumberVar ArrayCounter;
Local StringVar array Alphabet;
Alphabet:=MAKEArray("RDCR08201", "RDCR08301",
    "RDCR08401, "RDCR501");
ArrayCounter := ArrayCounter + 1;
Alphabet[ArrayCounter]
D.
Shared NumberVar ArrayCounter;
Shared StringVar array Alphabet;
Alphabet:=MAKEArray("RDCR08201", "RDCR08301",
    "RDCR08401, "RDCR501");
ArrayCounter := ArrayCounter + 1;
Alphabet[ArrayCounter]
E. if {?Enter Product Name} = "Ball Point Pen" then 'Exclusive Pens'

Q14. You have made changes to an existing Repository based Custom Function. Which of the following will apply to other users of the Custom Function?

A. The changes made to the Custom Function will be automatically replicated in all other formulas that use this function.
B. You must change this Custom Function and all individual formulas that use this function
C. You must cut and past the Custom Function into all applicable reports
D. By using the Extractor, all changes will be replicated across all reports automatically

Q15. You want to modify a report Custom Function. What should you do?

A. Choose Report | Formula Workshop | select Report Custom Functions Editor and make the required changes, save changes
B. Choose Report Options | Formula Workshop | select Custom Function Editor and make the required changes, save changes
C. Choose Options | Formula Workshop | select Custom Function Editor and make the required changes, save changes

D. Choose View |Report | Formula Workshop | select Custom
Function Editor and make the required changes, save changes

**Q16.   You want to view the Custom Function Properties, i.e.
Category, Author... What should you do?**

A. Right-click Custom File and select Toggle Properties Display
B. Right-click View\Custom Function and select Toggle Properties
Display
C. Right-click Custom Function and select Toggle Properties Display
D. Right-click File\Custom Function and select Toggle Properties
Display

**Q17.   You want users to have access to a descriptive help section
which will enable them to obtain help when using your
Custom Function. Where can this be located?**

A. Right-click the Custom Function, and click Help Text
B. From the Field Explorer, Right-click the Custom Function, Toggle
Property Display and click Help Text
C. Select Edit from the Menu Bar and select Help Text
D. Right-click the Custom Function, Toggle Property Display and
click Help Text

**Q18.   Based on the following Custom Function, identify the
replaceable arguments?**

Function (StringVar v1)
If v1 startswith 'Examhints-RDCR08201' then right (v1, 7) else
v1
A. v1
B. startswith
C. Right
D. Function

**Q19.   Which of the following reflects the field type used to
create a Custom Function?**

A. Name
B. Summary
C. Category
D. Repository

E. Author
F. Return Type
G. Help Text

Q20. Which of the following are Custom Function Properties? *(Multiple Answers)*

A. Name
B. Summary
C. Category
D. Repository
E. Author
F. Return Type
G. Help Text

Q21. Which of the following make up the Custom Function Arguments? *(Multiple Answers)*

A. Name
B. Type
C. Description
D. Default Values

Q22. Which of the following statements are true?

A. When the Display In Experts checkbox is unchecked the Custom Function will appear in the Formula Editor and Not in the Formula Expert
B. When the Display In Experts checkbox is checked the Custom Function will not appear in the Formula Editor but will appear in the Formula Expert
C. When the Display In Experts checkbox is checked the Custom Function will only appear if the Formula Expert is clicked

Q23. The text section of the Custom Function is disabled what does this indicate?

A. The Custom Function has been deleted
B. You do not have rights to change the Custom Function
C. The Custom Function has not been refreshed
D. You have not disconnected the Custom Function from the Repository to make the necessary changes

Q24. **You want to group a set of Custom Functions. Which of the following will you use to group the title?**

A. Repository
B. Category
C. Return Type
D. Author

Q25. **You try to add a Custom Function to the Repository but it fails. What could the problem be?**

A. You have not installed crystal Reports properly
B. You need patch SDECRY11
C. You do not have rights to update the Repository database
D. None of the above

Q26. **You are using the Formula Expert to create a formula, and you notice only one of the three Custom Functions you created appears under Report Custom Functions —Formula Expert. Which of the following reasons apply?**

A. The Display In Experts checkbox is checked in the Custom Function area, Right-click the Custom Function- Toggle Property Display -Display In Experts uncheck box
B. The Display In Experts checkbox is unchecked in the Custom Function area, Right-click the Custom Function- Toggle Property Display and click the Display In Experts check box
C. A Custom Function will never appear in within the Formula Expert
D. The Display In Experts checkbox is unchecked in the Custom Function area, Right-click the Custom Function- Toggle Property Display, and check the Display In Experts checkbox

Q27. **You want to create a Custom Function. What should you do?**

A. From the toolbar, choose Report | Formula Workshop| right-click report Custom Functions, select new, enter name of the new Custom Function and click the Use Editor button
B. From the toolbar, choose Formula Workshop| right-click report Custom Functions, select new, enter name of the new Custom Function and click the Use Editor button

C. From the toolbar, choose Database | Formula Workshop| right-click report Custom Functions, select new, enter name of the new Custom Function and click the Use Editor button

D. From the toolbar, choose View | Formula Workshop| right-click report Custom Functions, select new, enter name of the new Custom Function and click the Use Editor button

Q28. **The Custom Functions dialogue box enables the Designer to apply a description to the Custom Function. Which of the following also applies?**

A. The grey sections of the dialogue box cannot be changed within the dialogue box

B. The grey sections of the dialogue box can be changed within the dialogue box

C. The grey sections of the dialogue box cannot be changed within the format dialogue box

Q29. **You want to use Crystal Reports Financial custom formulas. Where can this be located?**

A. Formula Workshop - Formula Editor| Operators Tree

B. Formula Workshop - Formula Editor| Functions Tree

C. Formula Workshop - Formula Editor| Formula Tree

D. Repository Explorer

Q30. You have created a Custom Function and would like to add it to the Repository. Which of the following methods apply?

A. Highlight the Custom Function code and select Add to Repository from the Operators window, the Business Objects logon dialog box will appear, enter your login details and click ok

B. Right-click the Repository within the Formula Workshop and select Add to Repository, the Business Objects logon dialog box will appear, enter your login details and click ok

C. Highlight the Custom Function code and select Add to Repository from the Functions window, the Business Objects logon dialog box will appear, enter your login details and click ok

D. Highlight the Custom Function code and select Add to Repository from the Basic Syntax window, the Business Objects logon dialog box will appear, enter your login details and click ok

Q31. You try to edit a Custom Function but the Formula Text Window is greyed out and you cannot edit the Custom Function. Which reason applies? *(Multiple Answers)*

A. All Custom Function functions cannot be edited once created and saved

B. The Custom Function has been saved to the Repository

C. The Custom Functions has not been disconnected from the Repository

D. The Custom Function must be refreshed

Q32. Custom Functions saved to the Repository will appear under which of the following?

A. Repository Functions

B. Repository Operators

C. Custom Functions

D. Repository Custom Functions

Q33. A Custom Function which has NOT been saved to the Repository will appear under one of the following sections?

A. Report Functions

B. Report Operators

C. Report Custom Functions

D. Custom Function

Q34. How can you determine if a Custom Function has been saved to the Repository? *(Multiple Answers)*

A. The Custom Function will appear under Repository Custom Function if it is saved to the Repository

B. There will be a vertical line beside the Custom Function

C. The Formula Text Window will be grayed out

D. The Custom Function will be editable within the Formula Text Window

Hyperlink Reports

Q1. You want to provide users with the functionality to email sales representatives directly within your report. You have email addresses stored in the database table, which you have placed on the report. Which of the following methods apply?

A. Right-click the report and select Format Field, click the Format tab and select Current E-mail Field Value from the Hyperlink Type
B. Right-click the report and select Format Field, click the Report tab and select Current E-mail Field Value from the Hyperlink Type
C. Right-click the field and select Format Field, click the Hyperlink Tab and select Current E-mail Field Value as the Hyperlink Type
D. Right-click the report and select Format Field, click the Paragraph Tab and select Current E-mail Field Value from the Hyperlink Type

Q2. Which of the following are formats available within the Hyperlink tab of the Format Editor? *(Multiple Answers)*

A. A Website On The Internet
B. Current Website Field Value
C. An Email Address
D. A File
E. Current E-Mail Field Value

Q3. Which of the following are DHTML Viewer Only options within the Hyperlink Tab? *(Multiple Answers)*

A. Another Report Object
B. Report Part Drill-Down
C. Email indentation
D. Map Email Hyperlink

Q4. Only Static Hyperlinks can be created via Crystal Reports

A. True
B. False

Q5. **Only Dynamic Hyperlinks can be created via Crystal Reports**

    C. True

    D. False

Q6. **Client Email Addresses exists within your database, you want to give users the functionality to click on the clients email address to activate their email, which off the following apply?**

    A. Right-click the client email address and select format field – hyperlink tab and select Current Email Field-Value

    B. Right-click the client email address and select format field – hyperlink tab and select a File

    C. Right-click the client email address and select format field – hyperlink tab and select A Website On The Internet

    D. Right-click the client email address and select format field – hyperlink tab and select Current Website Field Value

Q7. **You have created the following formula: 'http://www.'+ {Client.CompanyName}+ '.com' which you have placed on your report, when click this formula does not direct you to the customers website, what additional action is required?**

    A. Right-click the field – select format field – hyperlink tab - Website Field Value

    B. Right-click the field – select format field – hyperlink tab - Field Value

    C. Right-click the field – select format field – hyperlink tab - Current Website Field Value

    D. Right-click the field – select format field – hyperlink tab - Value

ANSWERS

Answers:Chapter 1: Create a Basic Report

Connect To a Data Source

Q1.   Answer: D
Q2.   Answer: B
Q3.   Answer: C
Q4.   Answer: A
Q5.   Answer: ABCDEFGH
Q6.   Answer: ABCD
Q7.   Answer: ABCDEFGHI
Q8.   Answer: C
Q9.   Answer: B
Q10.   Answer: ABCDFG
Q11.   Answer: D
Q12.   Answer: A
Q13.   Answer: C
Q14.   Answer: D
Q15.   Answer: A
Q16.   Answer: ABC
Q17.   Answer: A
Q18.   Answer: ABCDEFGH
Q19.   Answer: A
Q20.   Answer: ABE
Q21.   Answer: AB
Q22.   Answer: AB
Q23.   Answer: ABCDEF
Q24.   Answer: ABCDE
Q25.   Answer: ABCDE
Q26.   Answer: A

Add Tables

Q1:   Answer: E
Q2:   Answer: A
Q3   Answer: C
Q4:   Answer: A
Q5:   Answer: A

Describe the Design Environment

Q1:   Answer: ABEFG
Q2:   Answer: FG
Q3:   Answer: F
Q4:   Answer: A
Q5:   Answer: D
Q6:   Answer: C
Q7:   Answer: B
Q8:   Answer: A
Q9:   Answer: A
Q10:   Answer: C
Q11:   Answer:A
Q12:   Answer:C
Q13:   Answer:B
Q14:   Answer:A
Q15:   Answer: ABCDEF
Q16:   Answer: ACDEF
Q17:   Answer: ABCDEF
Q18:   Answer: ABCDGH
Q19:   Answer: ACDEFGHI
Q20:   Answer: CEGHI
Q21:   Answer: ABC
Q22:   Answer: AB
Q23:   Answer: ABCD
Q24:   Answer: A

Describe the Design Environment and Creating a New Report

Q1   Answer: D
Q2   Answer: E
Q3.   Answer: ABCDE
Q4:   Answer: A
Q5   Answer: E
Q6   Answer: E
Q7   Answer: ABCDE
Q8   Answer: D
Q9   Answer: B

Q10   Answer: E
Q11   Answer: E
Q12   Answer: ABCDE
Q13   Answer: ABCDE
Q14   Answer: ACDEG
Q15:   Answer: AD
Q16:   Answer: ABCD
Q17:   Answer: AB
Q18:   Answer: A
Q19:   Answer: D
Q20:   Answer: EF
Q21:   Answer: ABCDFGHIJ
Q22:   Answer: B
Q23:   Answer: ABDE
Q24:   Answer: E
Q25:   Answer: ABE
Q26:   Answer: AC
Q27:   Answer: B
Q28:   Answer: AD
Q29:   Answer: B
Q30:   Answer:C
Q31:   Answer: D
Q32:   Answer: EG
Q33:   Answer: A
Q34:   Answer: B
Q35:   Answer: B
Q36:   Answer: B
Q37:   Answer: B
Q38:   Answer: ABC
Q39:   Answer: BC
Q40:   Answer: D
Q41:   Answer: BC
Q42:   Answer: D
Q43:   Answer: A
Q44:   Answer: C
Q45:   Answer: ABEFGHI
Q46:   Answer: BCFGH
Q47:   Answer: BCDF
Q48:   Answer: C

Describe the Design Environment and Creating a New Report

Q49:   Answer: D
Q50:   Answer: AD
Q51:   Answer: A
Q52:   Answer: C
Q53:   Answer: B
Q54:   Answer: ABFG
Q55:   Answer: B
Q56:   Answer: B
Q57:   Answer: AB
Q58:   Answer: D
Q59:   Answer: D
Q60:   Answer: D
Q61:   Answer: C
Q62:   Answer: B
Q63:   Answer: C
Q64:   Answer: BC
Q65:   Answer: B
Q66:   Answer: A
Q67:   Answer: C
Q68:   Answer: D
Q69:   Answers: B
Q70:   Answer: C
Q71:   Answer: D
Q72:   Answer: A
Q73:   Answer: ABCDE
Q74:   Answer: ABCDE
Q75:   Answer: A
Q76   Answer: ABCD
Q77:   Answer: ABC
Q78:   Answer: ABCDEFGHI
Q79:   Answer: ABCDE

Insert and Position Objects on a Report

Q1.   Answer: A
Q2.   Answer: ABCD
Q3.   Answer: ABCDEFG
Q4.   Answer: AB
Q5.   Answer: A
Q6.   Answer: A
Q7.   Answer: ABCDE
Q8.   Answer: ABCDEF
Q9.   Answer: A
Q10.   Answer: ABC
Q11.   Answer: BD
Q12.   Answer: BE

Preview and Save a Report

Q1.   Answer: ABCDE
Q2.   Answer: A
Q3.   Answer: C
Q4.   Answer: B
Q5.   Answer: D
Q6.   Answer: D
Q7.   Answer: CD
Q8.   Answer: A
Q9.   Answer: B
Q10.   Answer: C
Q11.   Answer: AB
Q12.   Answer: D
Q13.   Answer: ABCD
Q14.   Answer: D
Q15.   Answer: CD
Q16.   Answer: C
Q17.   Answer: A
Q18.   Answer: BD
Q19.   Answer: A
Q20.   Answer: B
Q21.   Answer: E
Q22.   Answer: C

Q23.   Answer: C
Q24.   Answer: B
Q35.   Answer: A
Q36.   Answer: ABD
Q37.   Answer: A
Q38.   Answer: A
Q39.   Answer: ABCD
Q40.   Answer: AD
Q41.   Answer: C
Q42.   Answer: BCD
Q43.   Answer: ABC
Q44.   Answer: AB

Apply Record Selection

Q1:   Answer: C
Q2:   Answer: AE
Q3:   Answer: C
Q4:   Answer: A
Q5:   Answer: D
Q6:   Answer: C
Q7:   Answer: C
Q8:   Answer: AC
Q9:   Answer: A
Q10:   Answer: C
Q11:   Answer: B
Q12:   Answer: CD
Q13:   Answer: B
Q14:   Answer: ABD

Apply Record Selection

Q15:   Answer: G
Q16:   Answer: B
Q17:   Answer: DE
Q18:   Answer: D
Q19:   Answer: B
Q20:   Answer: B
Q21:   Answer: ABC

Q22:   Answer: A
Q23:   Answer: A
Q24:   Answer: ABE
Q25:   Answer: D
Q26.   Answer: A
Q27.   Answer: D
Q28.   Answer: E
Q29.   Answer: B
Q30.   Answer: A
Q31.   Answer: A
Q32.   Answer: B
Q33.   Answer: AB
Q34.   Answer: D
Q35.   Answer: AB
Q36.   Answer: C
Q37.   Answer: C
Q38.   Answer: A
Q39.   Answer: AC
Q40.   Answer: A
Q41.   Answer: AC
Q42.   Answer: AC
Q43.   Answer: AB
Q43.   Answer: ABC

Organize Data in a Report

Q1:   Answer: A
Q2:   Answer: A
Q3:   Answer: D
Q4:   Answer: ABDEFHI
Q5:   Answer: A
Q6:   Answer: C
Q7:   Answer: AD
Q8:   Answer: B
Q9:   Answer: A
Q10:   Answer: A
Q11:   Answer: B
Q12:   Answer: D
Q13:   Answer: B

Q14:   Answer: C
Q15:   Answer: AD
Q16:   Answer: D
Q17:   Answer: BH

Organize Data in a Report

Q18:   Answer: D
Q19:   Answer: A
Q20:   Answer: FG
Q21:   Answer: A
Q22:   Answer: C
Q23:   Answer: A
Q24:   Answer: B
Q25:   Answer: BF
Q26:   Answer: C
Q27:   Answer: B
Q28:   Answer: C
Q29:   Answer: A
Q30:   Answer: C
Q31:   Answer: CD
Q32:   Answer: AD
Q33:   Answer: A
Q34:   Answer: A
Q35.   Answer: AC
Q36.   Answer:BD
Q37.   Answer: A
Q38:   Answer: B

Answers: Chapter 2 Customize And Format a Report

Format Objects

Q1.   Answer: CE
Q2.   Answer: BCD
Q3.   Answer: Fixed = B. Floating =A
Q4.   Answer: BC
Q5.   Answer: A
Q6.   Answer: A
Q7.   Answer: B
Q8.   Answer: B
Q9.   Answer: B
Q10.   Answer: C
Q11.   Answer: A
Q12.   Answer: C
Q13.   Answer: E
Q14.   Answer: AF
Q15.   Answer: B
Q16.   Answer: B
Q17.   Answer: D
Q18.   Answer: AE
Q19.   Answer: A
Q20.   Answer: E
Q21.   Answer: C
Q22.   Answer: B
Q23.   Answer: A
Q24.   Answer: D
Q25.   Answer: C
Q26.   Answer: BC
Q27.   Answer: D
Q28.   Answer: B
Q29.   Answer: ABCDE
Q30.   Answer: AB
Q31.   Answer: ABCDE
Q32.   Answer: A
Q33.   Answer: ACD
Q34.   Answer: AB

Q35.   Answer: ACE
Q36.   Answer: A
Q37.   Answer: B
Q38.   Answer: C
Q39.   Answer: B
Q40.   Answer: E
Q41.   Answer: C
Q42.   Answer: E
Q43.   Answer: C
Q44.   Answer: AE
Q45.   Answer: A
Q46.   Answer: C

Format Objects

Q47.   Answer: D
Q48.   Answer: A
Q49.   Answer: ABC
Q50.   Answer: ABCD
Q51.   Answer: A
Q52.   Answer: A
Q53.   Answer: C
Q54.   Answer: B
Q55.   Answer: A
Q56.   Answer: B
Q57.   Answer: D
Q58.   Answer: AE
Q59.   Answer: B
Q60.   Answer: D
Q61.   Answer: ABCDEFG
Q62.   Answer: A
Q63.   Answer: ABCDEFG
Q64.   Answer: A
Q65.   Answer: C
Q66.   Answer: D
Q67.   Answer: A
Q68:   Answer: CD
Q69:   Answer: AB
Q70.   Answer: C

Q71. Answer: A
Q72. Answer: B
Q73. Answer: B
Q74. Answer: ACD
Q75. Answer: ABCD
Q76. Answer: B
Q77. Answer: B
Q78. Answer: A
Q79. Answer: BD
Q80. Answer: C
Q81. Answer: A
Q82. Answer: A

Adding Graphical Elements

Q1. Answer: B
Q2. Answer: D
Q3. Answer: ABC
Q4. Answer: ABCEF
Q5. Answer: B
Q6. Answer: ABCD
Q7. Answer: ABCD
Q8. Answer: A
Q9. Answer: D
Q10. Answer: A
Q11. Answer: ABCDE

Apply Section Formatting

Q1. Answer: B
Q2. Answer: A
Q3. Answer: A
Q4. Answer: ABCDE
Q5. Answer: E
Q6. Answer: D
Q7. Answer: A
Q8. Answer: C
Q9. Answer: E
Q10. Answer: D

Q11.   Answer: BDFGHI
Q12.   Answer:ABE
Q13.   Answer: AB
Q14.   Answer: D
Q15.   Answer: ACD
Q16.   Answer: D
Q17.   Answer: A
Q18.   Answer: B
Q19.   Answer: AB
Q20.   Answer: A
Q21.   Answer: B
Q22.   Answer: C
Q23.   Answer: ABCDEF
Q24.   Answer: ABCDE
Q25.   Answer: B
Q26.   Answer: D
Q27.   Answer: B
Q28.   Answer: D

Create a Chart

Q1.   Answer: ABCEFG
Q2.   Answer: ABCD
Q3.   Answer: A
Q4.   Answer: C
Q5.   Answer: B
Q6.   Answer: ABCEG
Q7.   Answer: A
Q8.   Answer: B
Q9.   Answer: AE
Q10.   Answer: D
Q11.   Answer: A
Q12.   Answer: C
Q13.   Answer: ABCD
Q14.   Answer: B
Q15.   Answer: D
Q16.   Answer: B
Q17.   Answer: D
Q18.   Answer: E

Q19.   Answer: B
Q20.   Answer: B
Q21.   Answer: A
Q22.   Answer: C
Q23.   Answer: A

Create a Chart

Q24.   Answer: D
Q25.   Answer: A
Q26.   Answer: ABCD
Q27.   Answer: B
Q28.   Answer: A
Q29.   Answer: E
Q30.   Answer: A
Q31.   Answer: A
Q32.   Answer: ABCDEFH
Q33.   Answer: ABC
Q34.   Answer: ABCD
Q35.   Answer: C
Q36.   Answer: B
Q37.   Answer: AD
Q38.   Answer: D
Q39.   Answer: ABCD
Q40.   Answer: ABCDEF
Q41.   Answer: ABCD
Q42.   Anwers: A
Q43.   Answer. ABC
Q44.   Answer. A
Q45.   Answer. A
Q46.   Answer: ABCDEF
Q47.   Answer: ABC
Q48.   Answer: E
Q49.   Answer: AB

Apply Report Templates

Q1.   Answer: B
Q2.   Answer: A

Q3.   Answer: AB
Q4.   Answer: C
Q5.   Answer: D
Q6.   Answer: C
Q7.   Answer: D
Q8.   Answer: B
Q9.   Answer: AB.
Q10.   Answer: D
Q11.   Answer: D
Q12.   Answer: ABCE

Answers: Chapter 3 Create a Formula

Q1.   Answer: AC
Q2.   Answer: A
Q3.   Answer: B
Q4.   Answer: A
Q5.   Answer: CD
Q6.   Answer: AB
Q7.   Answer: ABE
Q8.   Answer: B
Q9.   Answer: E
Q10.   Answer: B
Q11.   Answer: C
Q12.   Answer: E
Q13.   Answer: F
Q14.   Answer: ABE
Q18.   Answer: B
Q19.   Answer: B
Q20.   Answer: BD
Q21.   Answer: A

Use Functions and Operators - Financial

Q1.   Answer: B
Q2.   Answer: B
Q3.   Answer: B
Q4.   Answer: A
Q5.   Answer: D
Q6.   Answer: E
Q7.   Answer: A
Q8.   Answer: AC
Q9.   Answer: C
Q10.   Answer: A
Q11.   Answer: B
Q12.   Answer: B
Q13.   Answer: B
Q14.   Answer: B
Q15.   Answer: E
Q16.   Answer: F
Q17.   Answer: A
Q18.   Answer: D

Q19.   Answer: C
Q20.   Answer: A
Q21.   Answer: A
Q22.   Answer: B
Q23.   Answer: F
Q24.   Answer: A
Q25.   Answer: A
Q26.   Answer: G
Q27.   Answer: B
Q28.   Answer: B

Use Functions and Operators - Financial

Q29.   Answer: A
Q30.   Answer: B
Q31.   Answer: D
Q32.   Answer: E
Q33.   Answer: D
Q34.   Answer: C
Q35.   Answer: A
Q36.   Answer: B
Q37.   Answer: A
Q38.   Answer: C
Q39.   Answer: A
Q40.   Answer: A
Q41.   Answer: E
Q42.   Answer: CDF
Q43.   Answer: ABC

Use Functions and Operators – Date Functions and Operators

Q1.   Answer: A
Q2.   Answer: D
Q3.   Answer: CDE
Q4.   Answer: ABD
Q5.   Answer: B
Q6.   Answer: C

Q7.   Answer: C
Q8.   Answer: A
Q9.   Answer: B
Q10.   Answer: D
Q11.   Answer: B
Q12.   Answer: D
Q13.   Answer: C
Q14.   Answer: ABCDEFGHIJ
Q15.   Answer: ADE
Q16.   Answer: D
Q17.   Answer: A
Q18.   Answer: E

Functions and Operators – Maths Functions and Formulas

Q1.   Answer: E
Q2.   Answer: B
Q3.   Answer: C
Q4.   Answer: D
Q5.   Answer: A
Q6.   Answer: A
Q7.   Answer: D
Q8.   Answer: C
Q9.   Answer: C
Q10.   Answer: C
Q11.   Answer: C
Q12.   Answer: AB
Q13.   Answer: C
Q14.   Answer: D
Q15.   Answer: D
Q16.   Answer: D
Q17.   Answer: C
Q18.   Answer: D
Q19.   Answer: CE
Q20.   Answer: B
Q21.   Answer: D
Q22.   Answer: D

Functions and Operators – Strings Functions and Formulas

Q1.   Answer: B
Q2.   Answer: E
Q3.   Answer: A
Q4.   Answer: B
Q5.   Answer: A
Q6.   Answer: C
Q7.   Answer: D
Q8.   Answer: E
Q9.   Answer: AEF
Q10. Answer: ADE
Q11. Answer: A
Q12. Answer: AB
Q13. Answer: C
Q14. Answer: C
Q15. Answer: A
Q16. Answer: ABCD
Q17. Answer: B
Q18. Answer: C
Q19. Answer: C
Q20. Answer: BEF
Q21. Answer: D
Q22. Answer: B
Q23. Answer: AB

Use Functions and Operators – SQL Expressions

Q1.   Answer: A
Q2.   Answer: B
Q3.   Answer: A
Q4.   Answer: ABCDE
Q5.   Answer: E
Q6.   Answer: A
Q7.   Answer: B
Q8.   Answer: B

Q9.   Answer: A
Q10.   Answer: AB

Use Control Structures

Q1.   Answer: DE
Q2.   Answer: ABD
Q3.   Answer: D
Q4.   Answer: C
Q5.   Answer: AB
Q6.   Answer: ABCDEFGH
Q7.   Answer: C
Q8.   Answer: A

Use Variables

Q1.   Answer: BEFGHI
Q2.   Answer: ABDEFGH
Q3.   Answer: B
Q4.   Answer: C
Q5.   Answer: A
Q6.   Answer: ABC

Use Arrays

Q1.   Answer: B
Q2.   Answer: B
Q3.   Answer: A
Q4.   Answer: B
Q5.   Answer: E
Q6.   Answer: A
Q7.   Answer: AB
Q8.   Answer: EF
Q9.   Answer: D
Q10.   Answer: A

Use Arrays

Q11.   Answer: A
Q12.   Answer: A
Q13.   Answer: A
Q14.   Answer: A
Q15.   Answer: A
Q16.   Answer: A
Q17.   Answer: A
Q18.   Answer: D
Q19.   Answer: A
Q20.   Answer: ABCDG
Q21.   Answer: B
Q22.   Answer: B
Q23.   Answer: A
Q24.   Answer:A

Answers: Chapter 4 Manage a Report

<u>Export a Report</u>

Q1. Answer: ABCD
Q2. Answer: G
Q3. Answer: ABCDE
Q4. Answer: A
Q5. Answer: C
Q6. Answer: BCD
Q7. Answer: ACD
Q8. Answer: D
Q9. Answer: A
Q10. Answer: C

<u>Manage Reports Using the Workbench</u>

Q1.   Answer: AD
Q2.   Answer: B
Q3.   Answer: ABC
Q4.   Answer: ABCD
Q5.   Answer: B
Q6.   Answer: A
Q7.   Answer: D
Q8.   Answer: A
Q9.   Answer: A
Q10.   Answer: A
Q11.   Answer: B

<u>Repository</u>

Q1.   Answer: C
Q2.   Answer: E
Q3.   Answer: BF
Q4.   Answer: D
Q5.   Answer: ABC
Q6.   Answer: ABC
Q7.   Answer:C
Q8.   Answer: C

Q9.   Answer: C
Q10.   Answer:ABDE
Q11.   Answer: A
Q12.   Answer: C
Q13.   Answer: DE
Q14.   Answer: A
Q15.   Answer: A
Q16.   Answer: A
Q17.   Answer: AB
Q18.   Answer: B

Repository

Q19.   Answer:A
Q20.   Answer: BE
Q21.   Answer: E
Q22.   Answer: B
Q23.   Answer:B
Q24.   Answer:ABC
Q25.   Answer: B
Q26.   Answer: D
Q27.   Answer:C
Q28.   Answer: B
Q29.   Answer: D

Answers: Chapter 5 Create an Advanced Report

Create a Parameter

Q1:   Answer: ABC
Q2:   Answer: A
Q3:   Answer: B
Q4:   Answer: B
Q5:   Answer: CD
Q6:   Answer: ACD
Q7:   Answer: D
Q8:   Answer: D
Q9:   Answer: D
Q10:   Answer: ABC
Q11:   Answer: B
Q12:   Answer: D
 Q13:   Answer: D
Q14:   Answer: C
Q15:   Answer: ABDE
Q16:   Answer: B
Q17:   Answer: B
Q18:   Answer: C
Q19:   Answer: ABC
Q20:   Answer: A
Q21:   Answer: B
Q22:   Answer: D
Q23:   Answer: A
Q24:   Answer: B
Q25:   Answer: B
Q26:   Answer: ABCE
Q27:   Answer: A
Q28:   Answer: C
Q29:   Answer: B
Q30:   Answer: A
Q31:   Answer: D
Q32:   Answer: ABCD
Q33:   Answer: C

Q34:   Answer B
Q35   Answer ABC
Q36.   Answer: A
Q37.   Answer: A
Q38.   Answer: A
Q39.   Answer: D
Q40.   Answer: A
Q41.   Answer: A

Create a Parameter

Q42.   Answer: A

Q43.   Answer: B

Use Dynamic Cascading Prompting

Q1.   Answer: A
Q2.   Answer: A
Q3.   Answer: B
Q4.   Answer: A
Q5.   Answer: A

Build And Format A Basic Cross-Tab

Q1. Answer: AD
Q2. Answer: D
Q3. Answer: AB
Q4. Answer: A
Q5. Answer: C
Q6. Answer: C
Q7. Answer: B
Q8. Answer: A
Q9. Answer: B
Q10. Answer: B
Q11. Answer: AB
Q12. Answer: A
Q13. Answer: ABDE
Q14. Answer: B
Q15. Answer: A
Q16. Answer: A

Build And Format A Basic Cross-Tab

Q17. Answer: A
Q18. Answer: D
Q19. Answer: A
Q20. Answer: A
Q21. Answer: A
Q22. Answer: CD
Q23. Answer: A
Q24. Answer: B
Q25. Answer: ABCDEF
Q26. Answer: C
Q27. Answer: C
Q28. Answer: A
Q29. Answer: A
Q30. Answer: B
Q31. Answer: ABCD
Q32. Answer: ABCDE
Q33. Answer: A
Q34. Answer: B

Q35. Answer: ABCEF
Q36. Answer: A
Q37. Answer: B
Q38. Answer: A
Q39. Answer: C
Q40. Answer: C

Use the Running Total Expert

Q1. Answer: B
Q2. Answer: A
Q3. Answer: B
Q4. Answer: B
Q5. Answer: ABCE
Q6. Answer: BCDF
Q7. Answer: A
Q8. Answer: D
Q9. Answer: A
Q10. Answer: D
Q11. Answer: ADE
Q12. Answer: AEDFG

Build a report With Alerts

Q1. Answer: B
Q2. Answer: B
Q3. Answer: AB
Q4. Answer: C
Q5. Answer: C
Q6. Answer: D
Q7. Answer: A
Q8. Answer: ABC

Build a report With Alerts

Q9. Answer: A
Q10. Answer: A
Q11. Answer: D
Q12. Answer: C

Build a Top N Report

Q1.   Answer: A
Q2.   Answer: C
Q3.   Answer: C
Q4.   Answer: ABCEG
Q5.   Answer: B
Q6.   Answer: A
Q7.   Answer: A

Answers: Chapter 6 USE REPORT PROCESSING TECHNIQUES

Explain the multi-pass reporting process

Q1.   Answer: ABCDG
Q2.   Answer: C
Q3.   Answer: B
Q4.   Answer: D
Q5.   Answer: A
Q6.   Answer: A
Q7.   Answer: C
Q8.   Answer: B
Q9.   Answer: D
Q10.   Answer: A
Q11.   Answer: A
Q12.   Answer: G
Q13.   Answer: E
Q14.   Answer: ABCDEFGHIJ
Q15.   Answer: C
Q16.   Answer: D
Q17.   Answer: G
Q18.   Answer: G
Q19.   Answer: B
Q20.   Answer: G
Q21.   Answer: G

Use the database expert

Q1:   Answer: A
Q2:   Answer: B
Q3:   Answer: AD
Q4:   Answer: E
Q5:   Answer: C
Q6:   Answer: D
Q7:   Answer: ABC
Q8:   Answer: D
Q9:   Answer: B

Q10:   Answer: DE
Q11:   Answer: A
Q12:   Answer: AB

Set Up and Configure Data Sources

Q1:   Answer: ACDFH
Q2:   Answer: B
Q3:   Answer:AC
Q4:   Answer: AB
Q5:   Answer: A
Q6:   Answer: A
Q7:   Answer: D
Q8:   Answer: AC
Q9:   Answer:ABCD
Q10:   Answer: A
Q11:   Answer: D
Q12:   Answer: B
Q13:   Answer: B
Q14:   Answer: C

Update Reports for Database Changes

Q1:   Answer: A
Q2:   Answer: A

Explain How to Validate Report Data and Validate Report Data

Q1:   Answer: D
Q2:   Answer: ABCDE

Distribute Personalized Content

Q1:   Answer: C
Q2:   Answer: ABCD
Q3:   Answer: C
Q4:   Answer: A

Q5:   Answer: A
Q6:   Answer: B
Q7:   Answer: B
Q8:   Answer: ABC

Use an Xml Transform

Q1:   Answer: A
Q2:   Answer: C
Q3:   Answer: D
Q4:   Answer: D
Q5:   Answer: A
Q6:   Answer: A
Q7:   Answer: B
Q8:   Answer: ABCD
Q9:   Answer: D

Use an Xml Transform

Q10:   Answer: A
Q11:   Answer: D
Q12:   Answer: ABC
Q13:   Answer: AB
Q14:   Answer: C
Q15:   Answer: A
Q16:   Answer: ABC
Q17:   Answer: A
Q18:   Answer: C
Q19:   Answer: B

Identify Links and Join Types

Q1:   Answer: D
Q2:   Answer: AB
Q3:   Answer: A
Q4:   Answer: ABC
Q5:   Answer: ACD
Q6:   Answer: C

Q7:   Answer: D
Q8:   Answer: B
Q9:   Answer: AD
Q10:   Answer: B
Q11:   Answer: ABCDGH
Q12:   Answer: A
Q13:   Answer: E
Q14:   Answer: F
Q15:   Answer: C
Q16:   Answer: B
Q17:   Answer: D
Q18:   Answer: C
Q19:   Answer: C
Q20:   Answer: C
Q21:   Answer: D
Q22:   Answer: B
Q23:   Answer: B
Q24:   Answer: B
Q25:   Answer: A
Q26:   Answer: B
Q27:   Answer: BCDE
Q28:   Answer: B
Q29:   Answer: E
Q30:   Answer: C
Q31:   Answer: D

Process Data on the Server

Q1:   Answer: ACEF
Q2:   Answer: AD
Q3:   Answer: ABCD
Q4:   Answer: ABCDE
Q5:   Answer: BEFG

Answers: Chapter 7 Use Subreports

Q1.   Answer: B
Q2.   Answer: E
Q3.   Answer: A
Q4.   Answer: A
Q5.   Answer: B
Q6.   Answer: D
Q7.   Answer: D
Q8.   Answer: A
Q9.   Answer: A
Q10.   Answer: B
Q11.   Answer: C
Q12.   Answer: D
Q13.   Answer: D
Q14   Answer: B
Q15.   Answer: B
Q16.   Answer: CE
Q17.   Answer: C
Q18.   Answer: BC
Q19.   Answer: C
Q20.   Answer: C
Q21.   Answer: A
Q22.   Answer: B
Q23.   Answer: ACDE
Q24.   Answer: ABCEG
Q25.   Answer: C
Q26.   Answer: D
Q27.   Answer: B
Q28.   Answer: C
Q29.   Answer: B
Q30.   Answer: D
Q31.   Answer: D
Q32.   Answer: B
Q33.   Answer: B
Q34.   Answer: C
Q35.   Answer: A
Q36.   Answer: C
Q37.   Answer: A

Q38.   Answer: A
Q39.   Answer: A
Q40.   Answer:B
Q41.   Answer: D
Q42.   Answer: ABDE
Q43   Answer: AB
Q44   Answer: A

Answers: Chapter 8 Create Complex Formulas

Use Evaluation Time Functions

Q1.   Answer: ABCD
Q2.   Answer: D
Q3.   Answer: D
Q4.   Answer: B
Q5.   Answer: ABC
Q6.   Answer: B
Q7.   Answer: B
Q8.   Answer: B
Q9.   Answer: E
Q10.   Answer: C

Use Print State Functions

Q1.   Answer: A
Q2.   Answer: ABC
Q3.   Answer: C
Q4.   Answer: B
Q5.   Answer: D
Q6.   Answer: A
Q7.   Answer: BC
Q8.   Answer: B

Use a Dynamic Array

Q1.   Answer: ABC
Q2.   Answer: A
Q3.   Answer: A
Q4.   Answer: B
Q5.   Answer: CDE

Use Loop Control Structures with Arrays

Q1.   Answer: AD
Q2.   Answer: D
Q3.   Answer: A

Index

VDB 152
verify 30
Verify Database 32

W

While loop 302
WhilePrintingRecords 249
WhileReadingRecords 249

X

XIRR 149
XML 8
XML and Web Services 8
XML export format 266
XML Transform 266
XNPV 150

Y

Year(CurrentDate) 57, 59
YearToDate 56, 59
Yield 153
YieldDisc 153
YieldMat 153

Z

zoom 46
zoom control box 46

Q4.   Answer: D
Q5.   Answer: ABCD
Q6.   Answer: B
Q7.   Answer: ABCDE

Use Custom Functions

Q1.   Answer: E
Q2.   Answer: B
Q3.   Answer: A
Q4.   Answer: C
Q5.   Answer: A
Q6.   Answer: B
Q7.   Answer: ABCD
Q8.   Answer: C
Q9.   Answer: B
Q10.   Answer: D
Q11.   Answer: B
Q12.   Answer: AB
Q13.   Answer: CE
Q14.   Answer: A
Q15.   Answer: A
Q16.   Answer: C
Q17.   Answer: D
Q18.   Answer: A
Q19.   Answer: F
Q20.   Answer: ABCDEFG
Q21.   Answer: ABCD
Q22.   Answer : A
Q23.   Answer: D
Q24.   Answer: B
Q25.   Answer: C
Q26.   Answer: D
Q27.   Answer: A
Q28.   Answer: A
Q29.   Answer: B
Q30.   Answer: B
Q31.   Answer: BC
Q32.   Answer: D

Q33.   Answer: C
Q34.   Answer: ABC

Hyperlink Reports

Q1.   Answer: C
Q2.   Answer: ABCDE
Q3.   Answer: AB
Q4.   Answer: B
Q5.   Answer: B
Q6   Answer: A
Q7.   Answer: C

About the Author

Antonia Iroko is the Director of Projection Programmers Ltd - a Business Intelligence Reporting Development Consulting Company which over the years had provided consulting and training services to financial, banking, manufacturing, pharmaceutical and fortune 500 companies both nationally and Internationally.

Antonia is the editor of the Examhints.com (http://www.examhints.com) which is one of the first websites to provide study guides for the Crystal Reports Certified Professional exams. Antonia has authored many customized training manuals for various companies, which are currently being used for user training. BOCP for Crystal Reports quick reference study guide was the first edition written by Antonia Iroko.

Antonia graduated from the University of Nottingham with a **BEng in Manufacturing Engineering and Operations Management** and then went on to Study for **an MSc in Computer Science**. She is a **Business Objects Certified Professional for Crystal Reports, OCP Oracle and MCP SQL Server**. Antonia now provides Business Intelligence consulting services to various financial, manufacturing and pharmaceutical companies. For consulting services visit http://www.projectionpro.com